Computer Graphics in Biology

ROBERT RANSOM and RAYMOND J. MATELA

Department of Biology,
The Open University

CROOM HELM
London & Sydney

DIOSCORIDES PRESS
Portland, Oregon

Croom Helm Ltd, Provident House, Burrell Row,
Beckenham, Kent BR3 1AT
Croom Helm Australia Pty Ltd, Suite 4, 6th Floor,
64–76 Kippax Street, Surry Hills, NSW 2010, Australia

British Library Cataloguing in Publication Data

Ransom, Robert
Computer graphics in biology.
1. Biology——Data processing 2. Computer
graphics
I. Title II. Matela, Raymond J.
574'.028'566 QH313
ISBN 0-7099-4106-4

First published in 1986 in the U.S.A. by
Dioscorides Press, 9999 S.W. Wilshire, Portland,
Oregon 97225, U.S.A.

ISBN 0-931146-05-4

Typeset in 10pt Times Roman by Leaper & Gard Ltd, Bristol, England

Printed and bound in Great Britain by
Biddles Ltd, Guildford and King's Lynn

Contents

Preface

Computer graphics is being used to an increasing extent in the biological disciplines. As hardware costs drop and technological developments introduce new graphics possibilities, researchers and teachers alike are becoming aware of the value of visual display methods.

In this book we introduce the basics of computer graphics from the standpoints of both hardware and software, and review the main areas within biology to which computer graphics have been applied. The computer graphics literature is vast, and we have not been able to give a full course on graphics techniques in these pages. We have instead tried to give a fairly balanced account of the use of graphics in biology, suitable for the reader with some elementary grounding in computer programming. We have included extensive references both to material cited in the text and to other relevant publications.

One of the factors that has fuelled the increase in graphics use is the ease with which the more simple graphics techniques may be implemented on microcomputers. We have, therefore, paid attention to microcomputer graphics as well as graphics techniques suitable for larger machines. Our examples range from simple two-dimensional graph plots to highly complex surface representations of molecules that require sophisticated graphics devices and mainframe computers on which to run.

The book is separated into two logical sections. The first part concentrates on general graphics techniques, giving an overview from which the reader will be able to refer to other more specialised texts as required. The more geometric aspects of graphics are based on mathematical principles, especially matrix algebra and descriptive geometry: relevant aspects of these methods are included here. We have tried to extract the essential mathematical elements necessary for an understanding of graphics principles, and have included suggestions for further reading in the more advanced techniques. Indeed, more complex graphics techniques require knowledge of physics, differential geometry and vector analysis, but we have not discussed these areas in the present text. The second part of the book covers specific biological applications in the areas of reconstruction, molecular modelling, image capture and simulations. These are the major areas in which computer graphics have been used in biology. Computer-aided reconstruction systems, especially reconstructions of serially sectioned material, are providing useful data on spatial interrelations in systems as diverse as the insect retina and the diatom chromosomal apparatus. We discuss a range of such applications in the present book. Some kind of image capture system is a necessary prerequisite to reconstructional analysis by computer, and we have included a brief review of some of the relevant concepts. More 'traditional', digitizer based input is being supplanted in many cases by video input, although these methodologies involve more sophisticated software to abstract the required information from the image input. Non-invasive reconstruction methods are also becoming of importance, and we briefly describe the concepts behind tomography, an X-ray based imaging system.

Spatial relationships are also the key element in the use of computer graphics in molecular modelling, and it is here that the most sophisticated graphics displays are used. Of particular topicality in this area is the use of graphics techniques to look at DNA:protein interactions in molecular genetics, although graphics displays have been used for some years to study molecular conformations, for example in drug design.

The final biological application area is that of simulation, where the accent is less on 'reconstructional realism' and more on representation of system dynamics. The days of purely 'numeric' simulation output are (thankfully) numbered, and many new and novel animation effects are being used to aid interpretation of simulation data. Our strategy has been to write this part of the book with two types of reader in mind. The first category is made up of those people who may wish to write graphics programs 'from scratch', and who will therefore need a thorough base in graphics techniques. The second class of reader may wish to use or modify existing graphics packages in his or her field of interest. We have therefore discussed the various application areas with both standpoints in mind, although the separation is of course far from clear cut.

Computer graphics is of relevance to researchers, teachers and students in all biological disciplines. The recent upsurge in interest in graphical man-machine interfaces, notably the WIMP (window, icon, mouse, pointer) system seen on the Apple Mackintosh and Commodore Amiga computers and in the Digital Research GEM operating environment have proven that even the most mundane computer operations are both simplified and speeded up by graphics techniques. We hope that the present book will provide a useful introduction to the use of computer graphics in biology, and are confident that the future will see more and more biologists resorting to graphical output of their data.

Acknowledgements

We have received help from a number of sources during the preparation of this book. We are indebted to the following people for permission to reproduce Figures.

Professor A. Van Dam (Figure 1.3); Apollo Computer Inc. (Figures 3.4–3.6); Dr P. Moens (Figures 8.2–8.3); Dr J.R. McIntosh (Figure 8.7); Dr P. Luther (Figures 9.1–9.3); Joyce-Loebl Ltd (Figure 9.4); Dr R. Gordon (Figures 9.5, 9.7 and 9.8); Dr W. Anderson (Figure 10.3); Dr G. Zientara (Figure 10.4); Dr H. Meinhardt (Figure 11.1); Dr W. Düchting (Figure 11.6).

We would also like to thank Dr P. Luther for sending us a preprint of his work, and Sue Brooks for helping with the PLOTALL and SIMPLEPLOT examples. We also thank Bob Fletterick and colleagues at the Department of Molecular Biophysics and Biochemistry, UCSF, for access to an Evans and Sutherland PictureSystem 2. Paul Gabbot (Department of Physiology, Oxford University) helped with the description of the application of the CELL program in Chapter 8.

1 An Introduction to Computer Graphics

1.1 THE BEGINNINGS OF COMPUTER GRAPHICS

In the early days of computing, computer graphics were a novelty rather than a necessity. Part of the reason for this was the very restricted amount of memory available on the first computers, but of course the graphics devices that could be used were few and far between. Early in the 1950s the Whirlwind computer was used to produce simple line drawings using a cathode ray tube display, and the SAGE air defence system also used CRT (cathode ray tube) displays on which the operators pointed at images using light pens. It was not until 1963 that computer graphics really came of age. In that year, the MIT student Ivan Sutherland published the results of his PhD thesis entitled *Sketchpad: A Man-Machine Graphical Communication System* (Sutherland, 1963). The fundamental ideas developed by Sutherland have formed the basis of much of the interactive computer graphics in use today.

In the early years, computer graphics hardware was very expensive, and graphics studies were limited to computer science departments and commercial organisations that had sufficient funds to purchase and develop suitable graphics devices. The emergence of the Tektronix 4010 storage tube CRT display in 1968 made graphics available to the mass market, and the appearance of monochrome raster CRTs in the early 1970s carried on this expansion of graphics availability. Graphics was still the province of the computer science professional, however, because graphics software was still in its infancy. It was not until the mid and late 1970s that mainframe graphics packages were widely introduced. The appearance of graphics libraries like the Tektronix Plot-10 Interactive Graphics Library (see Chapter 3) made graphics much more 'user friendly', albeit still at a high price.

The last five years have seen two major developments that have increased the use of computer graphics. The drastic reduction of hardware costs has now reached the point where a colour raster CRT display offering medium graphics resolution can be purchased for less than £1500. The microcomputer revolution has been even more influential: machines like the Commodore PET and the Apple II series led to increasing microcomputer sophistication that now allows colour graphics to be created on machine costing under £200. Modern microcomputers also feature BASIC interpreters with built-in graphics commands allowing even the complete novice to produce simple computer graphics with a minimum of effort. Although the more sophisticated graphics applications still require expensive hardware, computer graphics is now available to any biologist with even the most modest of research budgets.

1.2 WHAT IS COMPUTER GRAPHICS?

Computer graphics is the visual representation of the information encoded within a computer. Although normal text output on the computer screen is, in the strictest sense, 'graphics,' computer graphics really begins when graphs, histograms, pictures and animations are displayed.

As recently as 15 or even ten years ago, graphics systems were still in the main remarkably primitive. As we have seen in the previous section, reducing costs and increasing sophistication of the hardware available have made it possible to produce colour-shaded pictures of high definition for the cost of an ultracentrifuge or amino acid analyser. The potential for the biologist to use computer graphics is seemingly without limit, and the main purpose of this book is to give the background knowledge necessary for the student of the life sciences to harness the power of computer graphics. Besides the ability to handle graphics techniques it is also necessary to appreciate the areas in which graphics may aid biological research. Many examples of the use of graphics in biology are also to be found in these pages.

1.3 COMPUTER GRAPHICS AND BIOLOGY

The phrase 'visual representation of information coded within the computer' has already been used to define computer graphics. The usefulness of graphics techniques for the biologist is then concerned with visual output of information representing some biological event, process or object. This is very much a blanket definition, and we can break it down into three general categories for illustrative purposes.

Firstly, graphics can be used for the preparation of diagrams, figures, slides and visual aids such as overhead transparencies. All kinds of graphic techniques can be used to prepare material of this kind, from simple line drawings in monochrome to complex three-dimensional shaded images in colour. The value of computer graphics in such cases lies in the power of a visual image to impart information.

Next comes representation of experimental data as a direct part of the experimental results. Here, computer graphics is used in an interactive manner to aid the analysis process. Display of the results of a series of biochemical experiments as perhaps a dynamically changing graph, or the display of a set of histological sections as a three-dimensional reconstruction can give the experimenter a 'feel' for his material that many never come from poring over sheets of nine-figure numbers.

The final use of graphics lies in the area of simulation, and it is often here that the graphics maestro has the most scope for creativity. With the results obtained from experiments, the programmer is limited to a representation of a given set of data. Simulations are experiments carried out totally within the confines of a computer, and the amount and type of data is at the behest of the programmer. Much greater flexiblity in the use of graphics techniques is therefore available to the researcher. Some examples of the kind of flexibility may be found in Chapter 11 below.

1.4 THE ELEMENTS OF A COMPUTER GRAPHICS SYSTEM

As you progress through this book you will find many different kinds of graphics system, from complex hardware costing hundreds of thousands of dollars to cheap home computers; from huge graphics software packages running on large mainframe computers to a few graphics commands implemented in a simple Basic interpreter on a microcomputer. There are in fact common areas between all these different classes of hardware and software, so in this short section we will introduce these elements.

What do you need to do graphics work? Let us first consider hardware. In the next two chapters we will go into a great deal of detail about the various classes of graphic device, and of course the type of work that you can do depends to a large extent on the device at your disposal. In order to simplify matters we will decide on a problem. A graph is to be drawn. The graph is to be quite accurate, so the 'teletype' mode of presentation, where the standard character matrices are used to define the axes, labels and points (Figure 1.1), is just too crude.

We can break down the requirements of the graph into a number of definite items. First we need a mechanism to define our picture, perhaps by use of a pen in a plotter or by directing the electron beam in a VDU. Next we need to be able to harness this mechanism to draw straight lines between defined points on the display surface (the plotter or VDU). Given the basic hardware to carry out these operations, the software must be considered. Markers must be available to mark the points on the graph, and the points must be joined up by either straight lines, or by a fitted 'curve' between them. Text must also be available to label the graph, and it should be possible to place the text at any required position on the screen, with the option to display *y* axis text at right angles to the *x* axis text.

For the moment we can largely forget the actual hardware (which will be discussed in detail in the next chapter) and concentrate on the programming problem of drawing the graph. Let us assume that a set of data is available for a particular biological process (Figure 1.2). The first pieces of information we need to know are the dimensions of the display surface, ana-

Figure 1.1: Example of a Graph Drawn Using Standard Teletypewriter Output

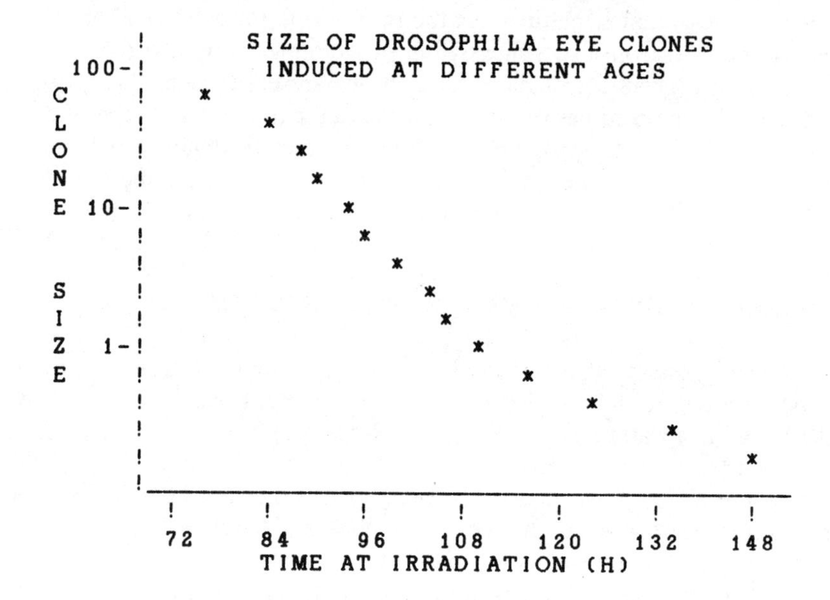

logous to the dimensions of a piece of graph paper. We will define these dimensions in 'display units', say $x = 640$, $y = 400$. Most computer graphic systems work with the origin (that is, point 0,0) of the coordinates at the bottom left corner of the display surface. The dimensions enable us to define both sizes and positions of the graph axes. If both x and y axes are to be the same length we might choose a size of 200 units with the origin at $x = 50$, $y = 50$.

The various elements drawn on the display surface — points, lines, circles, polygons and text characters — are termed graphics primitives. The variety of primitives that can be directly drawn by the software available is a function of the complexity of the graphics software available, and for the moment we will think only in terms of lines and character strings.

Our first programming task to draw a graph is to set up the graphics screen and to clear it of any existing data. The method used to do these operations will differ from system to system. On a small microcomputer like the IBM PC using interpreted BASIC, for example, a command of the form:

```
SCREEN 2
```

might be used to select the graphics screen instead of the text screen, and

Figure 1.2: Example of a Graph Drawn Using a Graphics Plotter

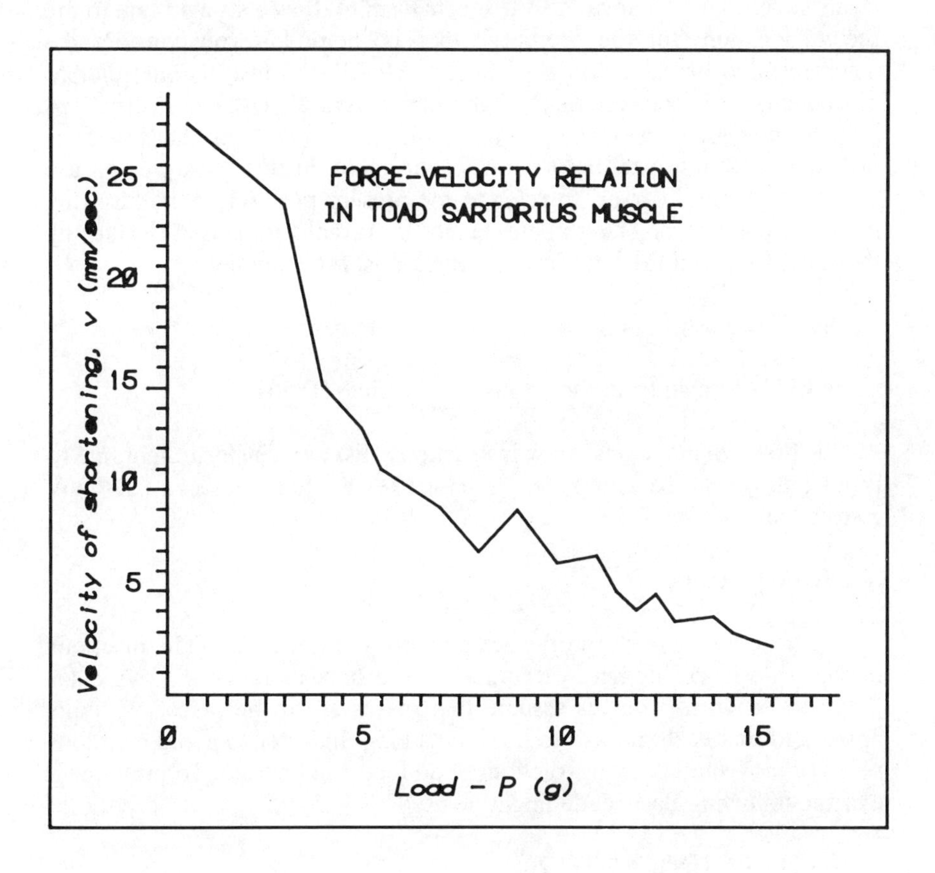

the screen might be cleared using a command of the form:

```
CLS
```

You will find some examples of graphics commands on various systems discussed more fully in Chapter 3, but for the moment we will restrict ourselves to a limited set of graphics commands of which versions are available in most graphics systems.

Once the screen has been set up and cleared, we need to draw the graph axes. For this we need some drawing commands that can be embedded into the graphing program. These commands are usually of the following form:

```
MOVE(50,250)
DRAW(50,50)
```

Where MOVE instructs the drawing mechanism to move to the given location, and DRAW instructs the mechanism to draw a straight line to the second location (for the present it may be helpful if you consider the mechanism to be a pen on a plotter, so all MOVE instructions refer to movements with the pen clear of the paper, and all DRAW instructions refer to movements with the pen touching the paper). Note that the *x* coordinate is always given before the *y* coordinate. In this book we will use the term 'graphics cursor' to refer to the present position of the 'drawing point' on the screen. The graphics cursor is normally initialised to (0,0) by the computer, so if the first draw instruction is, for example:

```
DRAW(50,50)
```

a line will be drawn from the origin to the point (50,50).

Our first two move and draw instructions give us a single straight line to begin our graph. To complete the bare axes we need a second DRAW instruction:

```
DRAW(250,50)
```

The next step is to put in the scale marks along the axes. The positions of the scale marks depend of course on the data used for the particular graph to be drawn. Let us assume that marks are to be placed 10 units apart, and that each mark is to be 5 units long. In order to avoid a tedious series of move and draw instructions, a do loop can be used. To mark the *x* axis the following Basic code might be used.

```
10 FOR COUNT = 1 TO 20
20        XP = 50 + (COUNT * 10)
30        MOVE(XP,40)
40        DRAW(XP,50)
50        NEXT COUNT
```

The similar code to mark the *y* axis would be as follows.

```
60 FOR COUNT = 1 TO 20
70        YP = 50 + (COUNT * 10)
80        MOVE(40,YP)
90        DRAW(50,YP)
100       NEXT COUNT
```

If two or more lengths of mark are required, a slightly more complex program segment would be required.

The next operation might be to place the text for the scale units and labels on the graph. Most simple graphics systems have the facility to place text on the graphics screen, and the normal convention is to place the text in a left-to-right horizontal arrangement from a chosen starting position. Text placement is one of the most irritating chores of the graphics programmer. The size of each graphics character cell in display units is predefined, so that the length of each text string can be found easily. The x,y start and finish points of the text string then have to be calculated with respect to the picture segments to be labelled. The y coordinate also has to take the height of the text cells into account.

More advanced text systems allow the variation of text size and orientation on the screen. In the most complex situations, text can even be projected onto a 'billboard' in three-dimensional space. These enhancements only add to the difficulty of choosing the correct start location for the text on screen. Fortunately, some tricks can be used, and the algorithm below illustrates the method used to label the x and y axes of our simple graph so that the text is always centred between the origin and maximum values on each axis.

Length of x string = number of characters × width of single character
Start coordinate for x axis text $= ((x^{\max} - x^{\min}) +$ length of string$)/2$

The y coordinate for the x axis text is chosen to give a suitable gap between the scale marks and the top of the characters, and with this we have the x,y start coordinate complete. A similar method is used to calculate the start position of the y axis text, but note one important exception: the text in this case must extend vertically instead of horizontally, and most graphics instruction sets possess the ability to perform text rotations at least along the vertical axis.

A typical statement for displaying text would be of the form:

```
MOVE(xp,yp)
TEXT(n,'CONCENTRATION')
```

where the text string 'CONCENTRATION' of length n characters is placed on the screen at the point (xp,yp).

The final component of the graph program would be the placement of the data points on the graph. In the simplest case, a series of (x,y) coordinates are read from a data file and each is plotted by placing a marker character at the (x,y) point on the graph. Although this could be done using single characters in a TEXT statement, special markers (triangles, circles, squares and so on) are often accessed by using a special statement, for example

MARKER(xp,yp,MT)

where a marker coded MT is placed at point (xp,yp).

The points on the graph could then be joined if required using a sequence of MOVE and DRAW commands in combination with the (x,y) data.

This simple outline of the graphic elements of a particular application demonstrates the features present in even the simplest of graphics packages or 'graphics enhancements' to high level language implementations. Even the most rudimentary graphics command sets will contain other features, and a method for clearing the screen is a universal requirement. The sorts of commands available in sophisticated graphics systems will be discussed in later chapters of this book.

If you have done much computing in the past you will realise that the best way to learn about computer techniques is to use them in an interactive manner. This is certainly true of computer graphics, and graphic methods are best learnt by tailoring them for your own applications. The basic armoury of graphics techniques can be mastered on any computer system capable of displaying the simple graph program outlined in this section, and it is only applications involving the generation of complex images that need more than a rudimentary graphics system. You are urged therefore to practise the construction and manipulation of graphics images as you read through this book, especially as you work through the central chapters on graphics methodologies.

1.5 COMPUTER GRAPHICS IN PERSPECTIVE

The basic function of a graphics program is to abstract elements from an object or set of objects and to represent these elements by computer display. Graphics books often term the objects to be represented in their environment as the 'world' and the programmer's job is therefore to construct a view of a picture of that world (Foley and Van Dam, 1982). As Foley and Van Dam put it, a graphics system may be considered as a 'synthetic camera'. A conceptual model of a graphics programming task might appear as in Figure 1.3.

As you can see, the important components of this model are the data structures, program and graphics system. Although this model looks very abstract, you will find yourself at the mercy of the tools available to you: for example use of a Fortran-based graphics package will limit the choice of data structures to arrays, while the Pascal programmer has a richer fund of flow of control constructs and data types to call upon. It is important to understand the capabilities of the computer system used before embarking on a full-scale application program.

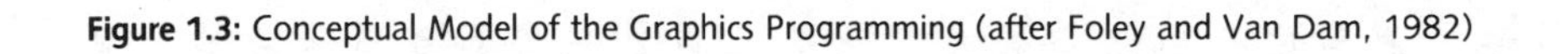
Figure 1.3: Conceptual Model of the Graphics Programming (after Foley and Van Dam, 1982)

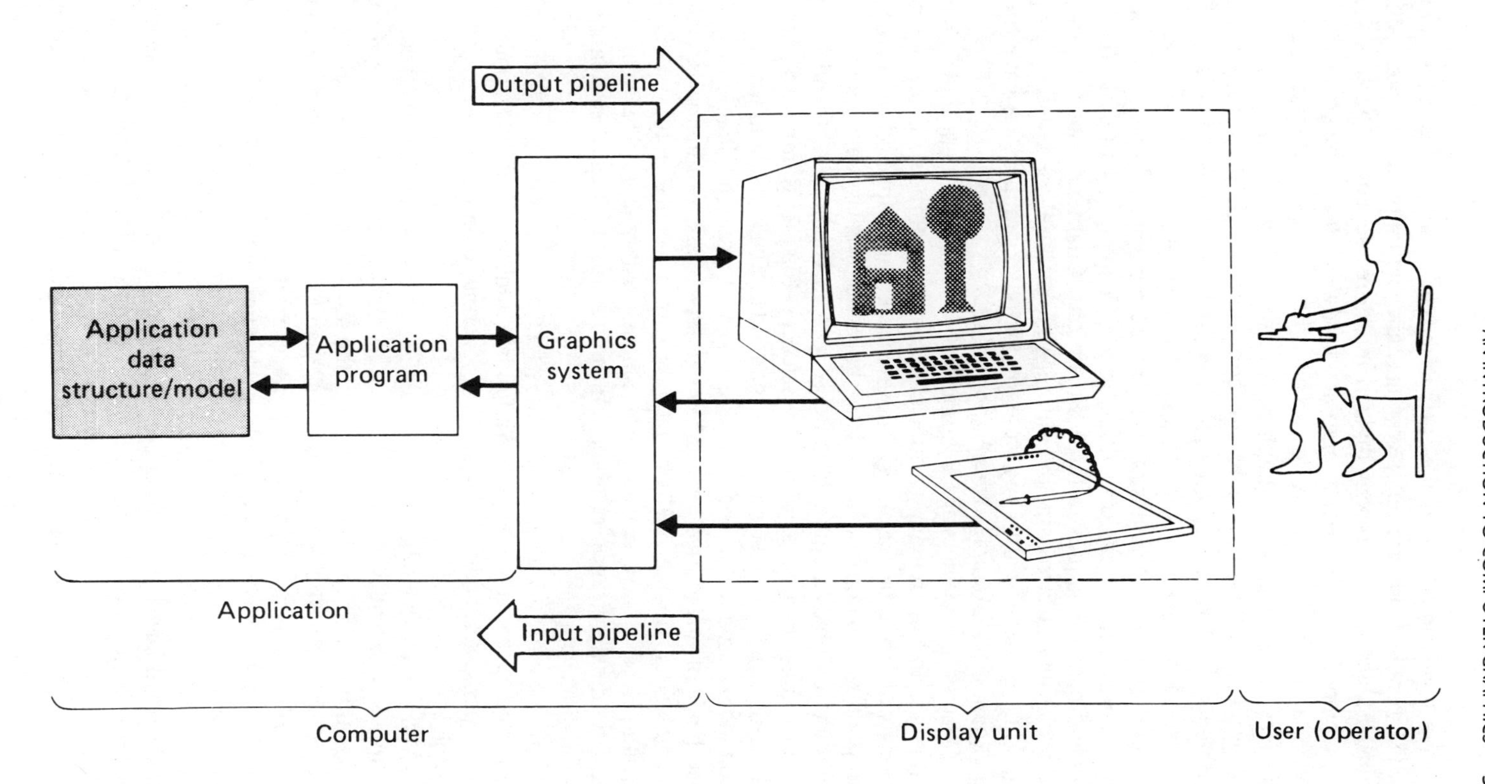

We have already seen the sorts of broad application areas in biology which benefit from computer graphics, and the following chapters are laid out as follows. The book is divided into three sections. The first section (Chapters 2 and 3) covers graphics hardware and software. Chapter 2 introduces the basic features of graphics hardware, from input and output devices to computers and graphic processors. Chapter 3 surveys the range of graphics software available on computers ranging from small micro-computers to large mainframe graphics packages.

The second section is made up of Chapters 4, 5 and 6, and deals with general graphic techniques of value to the biologist. Chapter 4 introduces two-dimensional transformations, and Chapter 5 extends the two-dimensional techniques to three dimensions. Chapter 6 looks at problems of visualising surface data, and although the material presented in Chapters 4, 5 and 6 is essential to any study of computer graphics, we have tried to use 'biological' examples where appropriate.

Chapter 7 considers the relationship of biological data and computer graphics, and the remaining three chapters cover three specific application areas within the life sciences. These chapters attempt both to review the sorts of applications developed to study biological systems, and to show how the graphics methodologies described in Chapters 4, 5 and 6 can be applied in specific areas of the life sciences. Chapter 8 introduces two and three-dimensional reconstruction methods. These techniques are most commonly used in the analysis of histological material. Chapter 9 looks at the fields of image capture and tomography — areas of increasing interest to biologists. Chapter 10 is concerned with molecular graphics, emphasis being placed on a review of graphics packages available. The final chapter reviews the application of graphic techniques for biological simulation and animation.

We have included two appendices for reference purposes. Appendix 1 revises matrix algebra, as this is a fundamental tool used in graphics transformations. Appendix 2 is a glossary of terms commonly used in computer graphics.

Many books have been written dealing with general aspects of computer graphics, and you will find reference to texts of this type throughout the present book. At the time of writing, the most complete and useful book is probably Foley and Van Dam (1982), and as this 'general' book runs to 660 pages you may well ask how we can attempt to give both a general introduction to computer graphics together with a study of the use of graphics in biology in a book less than a third of the size! The answer to this seemingly embarrassing question is that the Foley and Van Dam book was written largely with computer professionals in mind. The present book is designed for the biologist who wishes to master the rudiments of graphics techniques and to see how they may be applied in the life sciences. For these purposes we have not found it necessary to use a 'complete' approach; for example the chapter on three-dimensional transformations does not exhaustively cover every possible type of manipulation that can be performed on three-dimensional data. Our approach instead has been to cover the basic principles involved, so that the reader will be able to refer to other graphics texts in the most fruitful way.

Throughout this book we assume that the reader has at least a rudimentary knowledge of computer programming and has used some form of computer system. Much of the information here is however accessible to the computer-shy reader. A general introduction to computer use in the life sciences is given in Ransom and Matela (1985). Most computer users are familiar with at least the elementary features of several computer languages. We have kept the program examples to a minimum, and examples given here are in Fortran, Basic or Pascal.

1.6 REFERENCES

The classic book by Foley and Van Dam has already been recommended earlier in this chapter, and serious students of graphics will find it useful to get access to this book, especially as you read through the later chapters here. In many ways the forerunner to Foley and Van Dam (Newman and Sproull, 1979) is a gentler introduction to fundamental graphics techniques, but this book is now a little dated. Angell's slimmer volume (Angell, 1983) is an excellent general introduction to the geometrical side of computer graphics and contains many example subroutines written in Fortran. A very simple and readable introduction to microcomputer graphics is given by Hearn and Baker (1983). Sutherland's thesis is referenced mainly for completeness. The *Scientific American* article abstracts most of Sutherland's ideas and is entertaining to read.

Angell, I.O. *Introduction to Computer Graphics* (Macmillan, London, 1983)
Foley, J.D. and Van Dam, A. *Fundamentals of Interactive Computer Graphics* (Addison-Wesley, Reading, Massachusetts, 1982)
Hearn, D. and Baker, M.P. *Microcomputer Graphics* (Prentice Hall, Englewood Cliffs, New Jersey, 1983)
Newman, W.M. and Sproull, R.F. *Principles of Interactive Computer Graphics*, 2nd edn (McGraw-Hill, New York, 1979)
Ransom, R. and Matela, R.J. *Computers in Biology: An Introduction* (Open University Press, Milton Keynes, UK, 1985)
Sutherland, I.E. SKETCHPAD: *A Man–Machine Graphical Communication System* (Spartan Books, Baltimore, Maryland, 1963)
— 'Computer Inputs and Outputs', *Scientific American* (September 1966)

2 Graphics Hardware

2.1 AN OVERVIEW

Clearly, the nature of the graphical images that may be produced in a biological context is a function of the hardware and software available. These two components are closely linked, but for the purpose of clarity we will consider them as separate elements. In this chapter we concentrate on graphics hardware.

Any graphics system has four main hardware components, and these may be listed as:

Input devices
Display devices
Display processor
Main computer (that is, processor not dedicated to graphics).

The distinction between these subunits sometimes becomes blurred, for example microcomputers typically have display processors that are a single chip within the main computer housing, and input and output can both be done on a combined plotter/digitizer. A cathode ray tube (CRT) can also be used as an input as well as an output device, if joystick, light pen or mouse input are employed.

It is not really necessary for the biologist to appreciate the details of the physics and electronics of graphics devices, but a basic understanding of how the devices function is necessary if informed judgement of their capabilities is to be made. Different graphics hardware is suited to different tasks, and comparisons of drawing speed, resolution and colour capability require some basic knowledge. It is also important to appreciate the full range of hardware at the programmer's disposal. We will consider each of the above component categories in turn.

2.2 INPUT DEVICES

Input devices are used to transfer data into the computer. The most common input device is the keyboard, and although alphanumeric input via keyboard is really outside the definition of a graphics system component, it is possible to use the keys to control movement of a graphics cursor on a CRT, either by using 'cursor keys' if provided, or by assigning the input values returned by chosen keys in a program so that the movement of the graphics cursor can be controlled. Some keyboards designed for use with graphics CRTs have two 'thumbwheels' at right angles to each other (for example the ubiquitous Tektronix 4010 graphics terminal). Rotation of the thumbwheels gives cursor movement along both x and y axes.

The other input devices are designed specifically for graphics use. They include the light pen, touch-sensitive screen, mouse, tablet (or digitizer), joystick and image capture devices.

Light Pen

As you will probably know, the light pen is not a pen as such — in fact it does not usually emit light at all. It senses light from the CRT screen. Light pens contain a circuit board and a phototransistor or light-sensitive cell. The cell is sensitive to the high level of fluorescent light emitted when the electron beam hits the phosphor coating on a CRT screen. An electrical current is generated from the pen to the computer whenever the beam traverses the pen point. The pen is synchronised with the display processor in such a way that the x and y coordinates of the raster scan's (see below, p. 00) current position are returned for use by the programmer. Light pens are usually fitted with a switch to activate or deactivate the sensor.

The light pen is rather inaccurate as a locational device, and in this sense it is inferior to the tablet or joystick unless the degree of tolerance required is not high. One of the most useful functions of the light pen in past years has been to select options from a menu displayed on screen. The recent fashion for touch-sensitive screens and the mouse is reducing the instances in which the light pen is a truly useful graphics input device.

Touch-sensitive Screens

As you might expect, a touch-sensitive screen cannot be very accurate as a graphics input device due to the area of the human finger tip! A more apt name for a touch-sensitive screen is a touch panel, as the sensitivity is provided by a transparent panel mounted on the front of the CRT. Several methods have been used to sense the coordinates of the touch area. One of the most popular has been the use of two separated layers of transparent material, one of which has a film of conductive material over it, while the other layer is resistive. When a touch occurs, the two layers meet and a voltage drop occurs at that region. The coordinates of the point can be obtained from this voltage drop.

Mouse

The mouse is a device with one or more rollers on its base. If the mouse is moved over a flat surface the rollers turn, and potentiometers linked to these rollers sense the movements in each direction: a screen cursor is then moved to the appropriate x,y coordinates. Of course, if the mouse is lifted from the surface, the rollers will not turn and the screen cursor will remain at a given location. The mouse may also have one or more pushbuttons which may be used to input commands to the computer.

The mouse is considerably more precise than the light pen for input of coordinate data, but practice is needed if fine drawing is to be performed on a CRT. The mouse is a superlative device for menu-driven applications, especially if a system of icons (that is, symbols representing commands) is used on the CRT. This type of application has come to prominence on the Apple Mackintosh, and is now widely used in applications like word processing and database management.

Joystick

The joystick is a device that can be moved in one of several directions sensed by potentiometers. As with the mouse, it is difficult to control the fine movement of the screen cursor, and the most successful joystick applications are those in which direction rather than fine positioning is inputted from the joystick. Applications of this type are exemplified by games programs and flight simulators, where the joystick is used in much the same way as its namesake on an aeroplane is used to control pitch and roll. This device can perform the same functions as the mouse, and normally also has a pushbutton which can be used to input commands to the computer. It is also possible to obtain joysticks with a rotating knob to provide movement in the Z axis.

Digitizer

A digitizer is a device used for determining the coordinate data of an image. There are several types available, and essentially they consist of a flat sensitized surface to which the image to be digitized is attached. A moveable cursor is also used. The surface of the digitizer normally has a grid of wires embedded into it, and electromagnetic coupling is used to calculate the cursor's x,y position on the digitizer. One or more buttons are provided on the tablet or hand cursor, and the x,y coordinates of the cursor are returned either automatically (stream mode) or whenever a button is pressed (single mode). It is normally possible to obtain a high degree of accuracy when using a digitizer, and the accuracy available can be measured from the resolution and size of the digitizer. A digitizer of 12″ × 12″ (30 × 30 cm) with plotting accuracy of 200 points/inch or centimetre gives a resolution of 2400 × 2400 points, sufficient for the majority of purposes. A constraint on the usefulness of a digitizer is not the resolution but the actual size of the tablet, and the importance of size is linked to the use to which the digitizer is to be put. Most digitizer-based biological applications involve the input of spatial information from a pre-existing sketch, photograph or other material. If the material is larger than the area of the digitizer it will need either to be photographically reduced, or the data must be inputted in sections to be later 'joined' within the computer.

Hand cursors typically have a number of buttons used to input status information or commands to the computer. Buttons may, for example be used to start or finish transmission of a stream of (x,y) coordinates to the computer, or each (x,y) pair of coordinates may have a status code associated with it; if the input consists of single points, the Ascii code for the button used to specify the point may be sent to the computer at the same time as the coordinates, so that distinctions can be made between classes of data, so

17.69	23.52	A
23.23	42.46	A
21.21	23.12	B

might be a set of data when the first two columns represent the x and y coordinates, and the third column represents a code associated with each pair.

A typical 'biological' use of a digitizer might be to input outlines or dimensions of cells in a photomicrograph (see Chapter 8), or to input distribution data from a map for some kind of numerical analysis within the computer.

Tablet

A tablet is in effect a low resolution digitizer. It also responds to the action of a hand-held cursor over the tablet surface and is capable of sending coordinate data to the host computer. A typical use of a tablet would be to select menu options by applying an appropriate overlay to the tablet surface. The tablet could then be used to interactively change items and to invoke the appropriate responses.

Image Capture Device

An image capture device offers some of the benefits of a digitizing tablet for less manual effort. The device is based around a video camera that produces an image of an object, drawing or photograph. This image is then encoded into a two-dimensional matrix of grey shades or colours for processing by a computer. Unlike the devices that have been discussed above, the usefulness of an image capture device is closely linked to the memory capabilities of the computer to which it is coupled. Storage of a 500 × 500 pixel TV frame in its entirety will need a minimum of 93 Kbytes to store an image as simple as an eight-colour picture, but this could be reduced to 31 Kbytes if a single colour outline is to be stored. As you will see in Chapter 9, purpose-built microcomputer-based image analysis equipment is available to the biologist. This equipment has a specially designed CPU and graphics processor for handling the large amounts of data sent from the camera. In general, image analysis producing full-colour images can only be performed on mini computers or mainframe computers.

2.3 DISPLAY DEVICES

The most visible component of any graphics system is usually the display device, and the most common such device is the CRT. Other devices in this category include printers, plotters, and liquid crystal displays.

Printers

Printers have of course been used to display output since the earliest days of computing, and although text-based output is beyond the scope of this book, artful use of the limited capabilities of alphanumeric printers can result in impressive computer graphics. We can subdivide printers into the following categories.

Alphanumeric Printers. These printers can only output characters, numbers or other symbols in a standard matrix of print positions, say 120 horizontal characters by 50 vertical lines per page. There are three ways in which graphics output can be obtained with such printers. First, text output can be formatted to produce flowcharts or flow diagrams. Next, two-dimensional pictures can be obtained using the rows to represent y increments and the columns to represent x increments (Figure 2.1). Because of the large character cells (that is, the fixed physical space occupied by each character), the larger the picture on the page, the greater the resolution obtained. Finally, 'overprinting' of character cells can be used to give a 'grey scale' effect. This technique has been used to great artistic effect with representations of photographs, but it does require a printer capable of the overprinting operation.

Alphanumeric printers may be line printers, teletypewriters, daisywheel printers or dot matrix printers.

Dot-addressable Matrix Printers. The advent of this class of printer has made true graphics output available for an extremely low cost. All dot matrix printers work by utilising a matrix of needles, most commonly nine vertical by one horizontal. Each needle is driven by an electromagnet which 'fires' the needle under an electrical impulse. When a needle fires it impacts a carbon ribbon against the paper in the printer. If all nine needles are fired at once, a vertical bar is printed. If a lesser number of needles is fired, various patterns are produced. Each character is made up of a horizontal series of movements of the print head (for example 5 to give a 9×5 character matrix), so that a capital 'T', for example, might be made by firing the top needle only in columns 1,2,4 and 5 but all needles in column 3 (Figure 2.2).

Figure 2.1: Map Drawn Using Line Printer Output

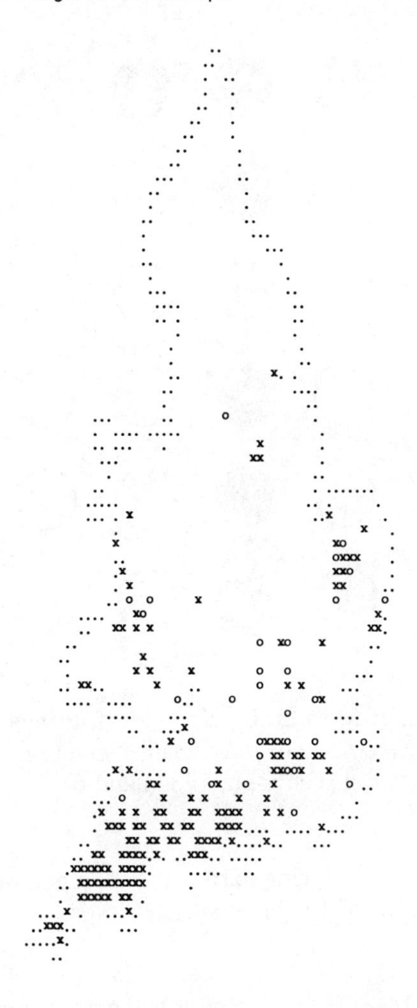

Dot-addressable matrix printers allow the user to access each needle individually. This property is used to produce a representation of the pixels on a graphics screen. A typical printer of this type is the EPSON MX80, which can address up to 576 dots per line, allowing the production of graphics output from a medium resolution graphics screen (say 640 × 480 pixels).

Ink Jet Printer. Ink jet printers work in a rather similar way to dot-addressable matrix printers, although a carbon ribbon and needles are not used to produce each dot. Instead, the ink jet printer 'shoots' ink of the three primary colours (red, green and blue) at the paper in the printer, the

Figure 2.2: Text Representation (Capital 'T') Using a Dot-addressable Matrix Printer

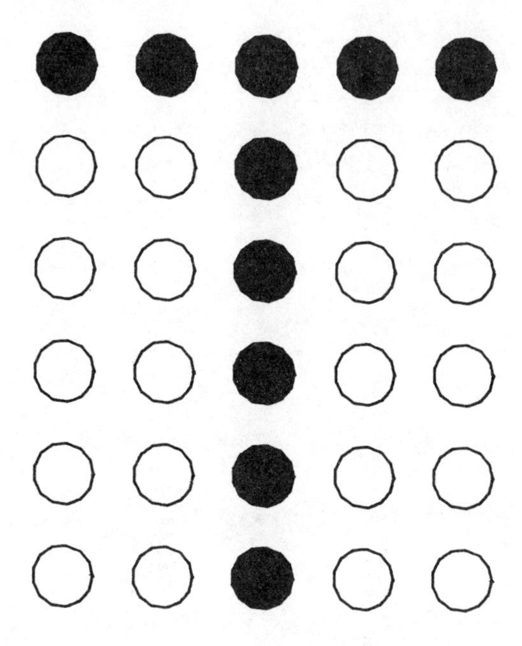

inks mixing on the paper to form the dot colours. Printers of this type are also becoming inexpensive, and their main limitation over dot matrix printers is speed of drawing (Harrison, 1984a and b).

Plotters

A plotter is a device for producing hard copies of a graphical image. From a technological point of view there are two types of plotter, pen and electrostatic.

Pen and Flatbed Plotters. The pen plotter is the most popular, and comes in two main forms, drum and flatbed. Drum plotters are used for large plots and tend to be expensive. They consist of a large rotating drum on which the paper or other drawing medium sits. The rotation of the drum forms one axis of the system. The pens usually ride on a carriage mounted on a long fixed arm that runs parallel to the long axis of the drum cylinder. This horizontal movement forms the other axis of the system.

The flatbed plotter consists of a flat square or rectangular stationary surface. The pens ride on a moveable carriage which sits on a moveable arm. These two movements form the two axes of the system. In both flatbed and drum plotters, the pens are usually interchangeable to provide wet ink, fibre point or ballpoint, and colour capability is only limited by the number of pens and inks available on the plotter. Multicolour output can even be produced on a single pen plotter if the colours are plotted sequentially and the pens are changed as needed.

As pen movement is controlled by an analogue system of servomotors and sensors, plotting a line involves drawing a genuine straight line rather than a series of stepped dots as with the dot matrix printer (Figure 2.3). Plotters are therefore useful for high quality output: figures for publication, posters and so on. Some older plotters (such as Tektronix 4663) incorporate a cursor on the pen head which can be moved using a joystick on the body of the plotter. Plotters of this type can therefore be used as digitizers, although the horizontal positioning of the plotter and the awkward position of the joystick and buttons on the body of the plotter make it a less than satisfactory input method for large amounts of graphic data.

Electrostatic Plotter. The final type of plotter that we will consider is the electrostatic plotter. These devices are usually expensive, if for no other reason than that they require special media. The image is produced by running the medium past a horizontal array of pins which, when fired place a charge at that location on the medium. The medium is then processed by wet chemical means, is dried, and is then ejected. Electrostatic plotters are also capable of producing high resolution colour images (Goddard, 1984).

Cathode Ray Tubes (CRTs)

Although we will consider CRTs and display processors in separate sections here, you should note that they are closely linked together. The

Figure 2.3: A Comparison Between Lines Drawn on (a) a Plotter, and (b) a Raster CRT. Note the characteristic 'staircase' effect produced on the raster device

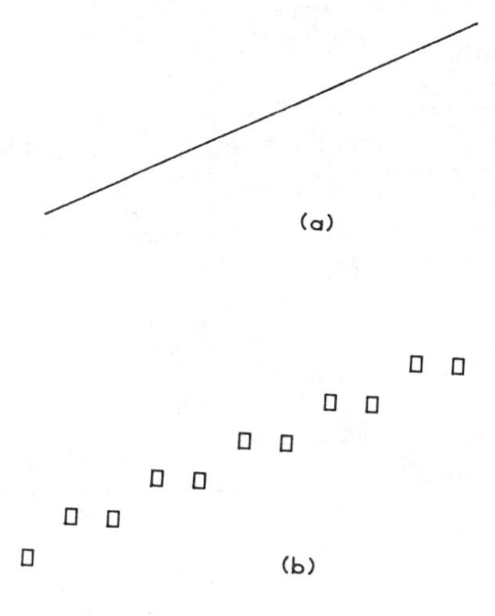

computer must have a method for rapidly encoding drawing instructions for the display device, and undoubtedly the most rapid form of graphic output is offered by the CRT. The general term 'CRT' covers several different classes of device that may be used for different purposes, from the rapid movements required in the display of animated or simulated images to extremely high resolution static pictures. It is important that the different technologies utilised by CRTs are appreciated in order to choose the correct device for a particular application or range of applications.

In general, a CRT consists of an electron gun which emits a stream of electrons towards a screen coated with a phosphorous compound (Figure 2.4). In order to direct the beam towards specific areas on the screen, two other components are needed. These are a focusing system and a method of beam deflection. Focusing is carried out by the induction of electrostatic or magnetic fields. The deflection system is the most important part of the CRT from a graphics point of view, because the speed at which the beam can be moved between points on the screen will govern the speed at which images can be drawn. Electrostatic or magnetic deflectors can be used, but most displays use magnetic systems because of their relative simplicity.

There are two major types of CRT, termed refresh and vector (or random scan) devices. In the former, the deflectors move the beam along

Figure 2.4: Structure of a Generalised CRT Device

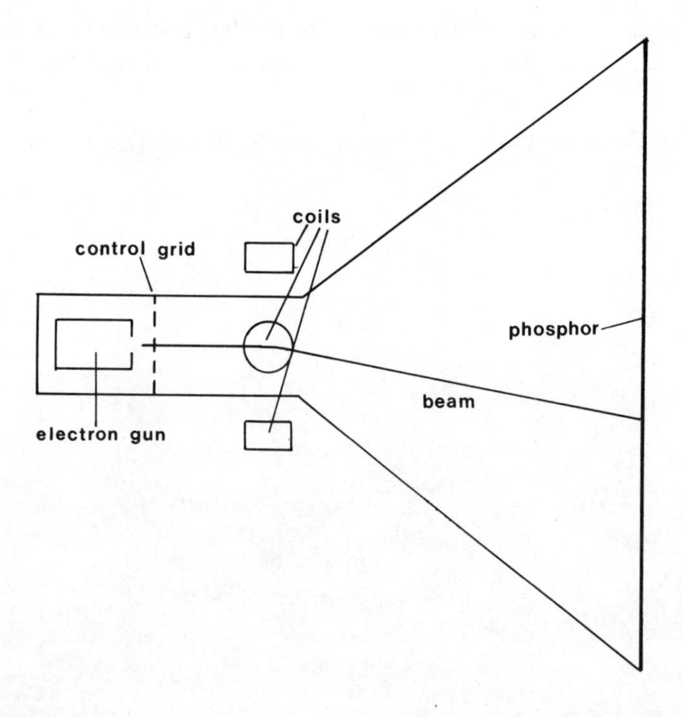

each row on the screen in turn, returning to the top as a refresh cycle is completed. The beam is intensified at the screen coordinate positions where a picture element is displayed, and each refresh cycle usually takes 1/60th of a second to complete. Random scan CRTs move the intensified beam only over the screen regions where picture elements are to be displayed, and the beam is blanked when movement is over areas not displayed. The beam path must be repeated before the phosphorescence decays. The advantage of this latter type of CRT is economy of beam movement: if only a few lines are drawn on the screen the picture can be displayed much more rapidly than with a refresh device, which has to scan every row on the screen to display an image.

The phosphor-coated screen is a further important component of the CRT. Whereas the deflectors determine the speed at which drawing takes place, the type of phosphor used determines the persistence of the image. Persistence is defined as the time taken for the phosphoresence to decay to 10 per cent of the initial light output, and is usually of the order of a few tens of microseconds (ms).

Storage Tube. In the vector devices, expensive memory and processors are required to maintain large lists and fast speeds. In response to these design characteristics the direct-view storage tube (DVST) was developed. The storage tube is a special class of CRT in which the beam does not have to be refreshed. The 'storage' ability is made possible by a storage surface and associated collector grid adjacent to the phosphor coat on the screen (Figure 2.5). When the electron beam hits the storage surface, the dislodged electrons are attracted to the collector grid. As the screen region that was hit then has a relative positive charge, and the storage surface does not conduct, the charge pattern is stored.

The image is written to the screen in the same way as with the random scan CRT, and erasure of the stored picture is performed by applying a positive voltage to the storage surface.

The advantage of the storage tube display is that it is a fairly cheap way of getting a high resolution flicker-free image, and one storage device, the Tektronix 4010, has become an industry graphics standard. The disadvantage of the storage tube is that selective erasure of parts of the image is not possible, and this makes storage tubes impossible to use for generation of animated images. This problem is compounded by the method of image erasure: when the positive voltage is applied to the screen, floodgun electrons are attracted to it, causing a flash of light. This makes it impossible even to generate smooth sequential series of whole images (for example rotation of a molecule).

Storage tubes are used when static, high-definition pictures are to be displayed, and for this they are ideal. The main advantage of the storage tube systems is their ability to produce 'clean' diagonal lines (that is lines

Figure 2.5: Structure of a Storage Tube CRT

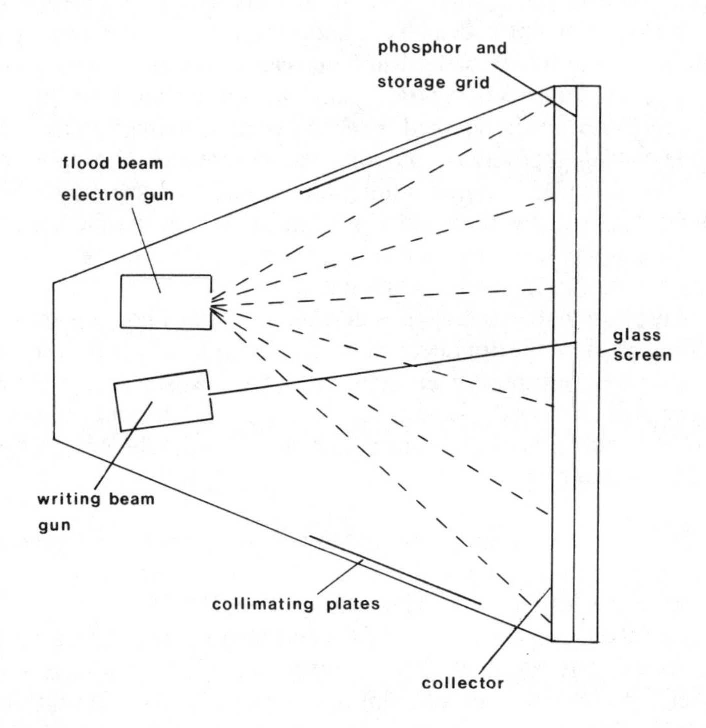

without the characteristic raster staircase effect) coupled with very fast drawing speeds. We mentioned above that the Tektronix 4010 graphics protocol has become an industry standard. The problem with this is that many refresh CRTs are now sold with the ability to emulate a 4010 device, and emulation of this type is limited to storage tube functions; refresh displays that are physically quite capable of selective erasure will not be able to carry it out when emulating a 4010. We will look in more detail at Tektronix graphics software in the next chapter.

CRT Resolution. With monochrome vector displays, the resolution is set by the limits of addressability of the deflectors used (Oakley, 1984), together with the size of the refresh buffer in the display processor (that is, the list of deflections needed to produce the displayed image when it is refreshed). Resolutions of 1024 × 1024 are common, and it is possible to go much higher than this. The limiting factor on monochrome refresh displays is the time needed to scan the screen (De'ath, Jones and Butcher, 1984). If the screen area is doubled, the scan time will also double, unless the beam deflection is speeded up. If a refresh rate of 1/60th of a second is needed to avoid 'flicker', the scan speed must be less than 1/60th of a second. At this refresh rate, images of 2048 × 1568 pixels can be displayed (for example, Westward 3219W graphic display) to give a flicker-free picture. However, refresh systems often use an interlace system. This means that even numbered raster scan lines are displayed in one 1/60th second scan, with odd numbered lines displayed in the next 1/60th second scan. Each scan is therefore only over half the full number of lines and takes half the time. This technique gets over the flicker that would occur if the whole screen were scanned at 1/30th of a second. The interlaced image does tend to look 'dull', however.

Colour CRTs have additional limits on resolution. Besides the complexity inherent in controlling three electron guns, one for each primary colour, the screen is made up of a matrix of phosphor dots arranged in triangles (Figure 2.6). Each triad contains one dot whose phosphor emits red when excited together with corresponding blue and green dots. A range of colours is produced from each triad by exciting each of the three dots to a different extent. In front of the screen is a 'shadow mask', a plate containing a hole for each triad, so that each dot in each triad is only exposed to the electrons from the 'correct' colour gun. In colour TV tubes, the triads are placed about 0.6 mm apart, so that a 12″ × 10″ (30 × 25 cm) screen has a maximum resolution of about 500 × 420 pixels. High resolution screens have triads as little as 0.35 mm apart giving almost twice this resolution. It is therefore possible to obtain a 1024 × 1024 colour raster display.

Figure 2.6: Representation of Colour on a CRT: Three Guns Are Used, One for Each Primary Colour (Red, Green, Blue, RGB)

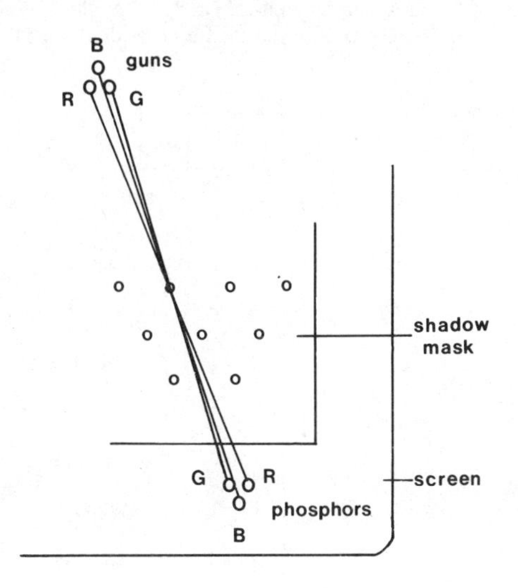

Colour Definition

As colour graphics displays are becoming more and more common, it is important to understand the principles of colour mixing and definition. The starting point for a discussion of colour is the hardware for colour generation. As we have seen, raster displays use the basic red, green and blue (RGB) method of producing colour. Another method of defining colour in a hardware device is used in many printing devices. These mix the colours cyan (blue), magenta (purple) and yellow to produce the desired colour, a method called the CMY system. Note that red + green + blue will equal white (additive primaries), whereas cyan + magenta + yellow will equal black (subtractive primaries).

The number of colours that can be displayed on a graphics device is a function of the number of levels of display memory (see below, p. 00) available, and if a single layer of memory is used, only one bit may be available for each of the three guns per pixel. This gives eight possible colour combinations depending on the on/off state of the red, green and blue guns (Figure 2.7). In this case the default colours will be red, green, blue, and white, together with their complements cyan, magenta, yellow and black. If the output of the individual guns can be varied, a number of substitute colours may be possible (but note that the number of displayable colours will still only be eight. The range of colour combinations is called the colour palette, and in the most expensive graphics terminals the palette can be up to 16.2 million colours. Each colour is hardware-defined as levels of red, green and blue, with the maximum as one part of each (yellow is defined as red: 1.0; green:1.0; and blue:0.0).

Figure 2.7: Specification of Colour Using One Bit for Each of the Primary Colours. This allows eight colours to be drawn, depending on the status of each of the three bits

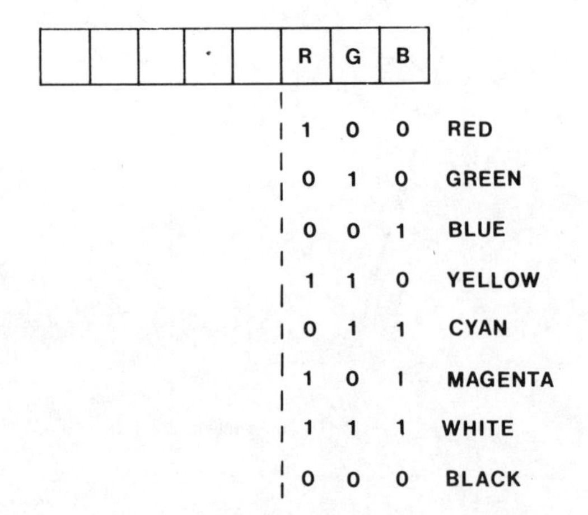

It is not easy to intuitively select colours by setting levels of red green and blue directly, as we tend to see colours using the 'artistic' concepts of hue, saturation and brightness. Two different 'colour models' have been developed to aid colour selection. These models are called the hue, lightness, saturation model (HLS system), and the hue, saturation, value model (HSV system). The models are best imagined as structures in three-dimensional space, the HSV system as a single cone, and the HLS system as a double cone. In both models, the primary 'pure' colours are found around the largest circumference of the cone: in the HSV system, for example, red is situated at 0 degrees, yellow is at 60 degrees, and cyan is at 300 degrees. With HSV, the saturation of the colours is highest at the circumference of widest part of the cone and diminishes towards the centre, while the value of the colours is highest at the centre of the widest part of the cone, and diminishes towards the cone's apex (Figure 2.8). With HLS, lightness is the parameter running through the central axis of the double cone, with black (0) at the lower point, and white (1) at the highest point (Figure 2.9). HLS saturation is defined in a similar way as in the HSV system. The HLS system is used by Tektronix for colour definition in the Plot 10 Interactive Graphics Library discussed in Chapter 3 below.

Figure 2.8: The HSV (Hue, Saturation, Value) Colour Model. Each colour is represented by values of the three parameters on a cone (here shown as a hexcone for simplicity). Note that the value parameter is 0.0 for black and 1.0 for white (compare Figure 2.9)

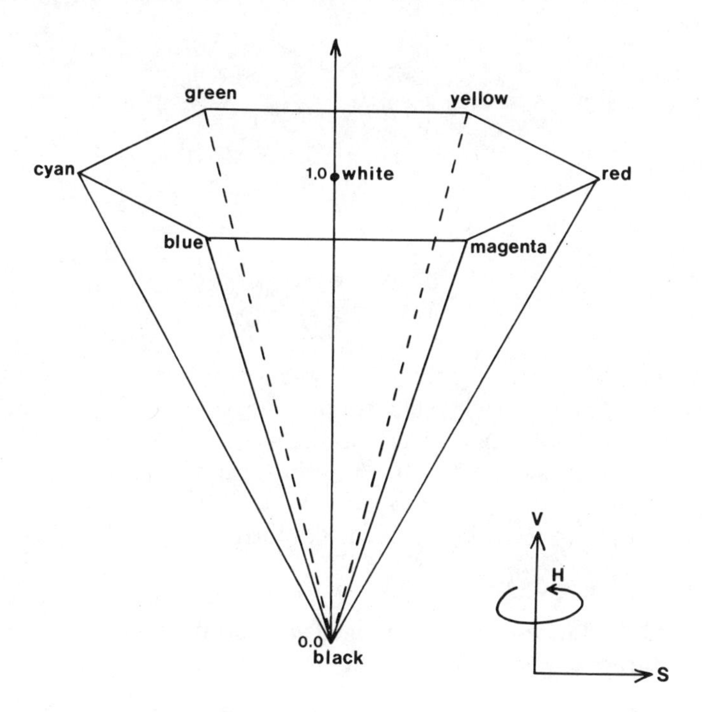

Figure 2.9: The HLS (Hue, Saturation, Lightness) Colour Model. Each colour is represented by values of the three parameters on a double cone (here shown as a hexcone for simplicity)

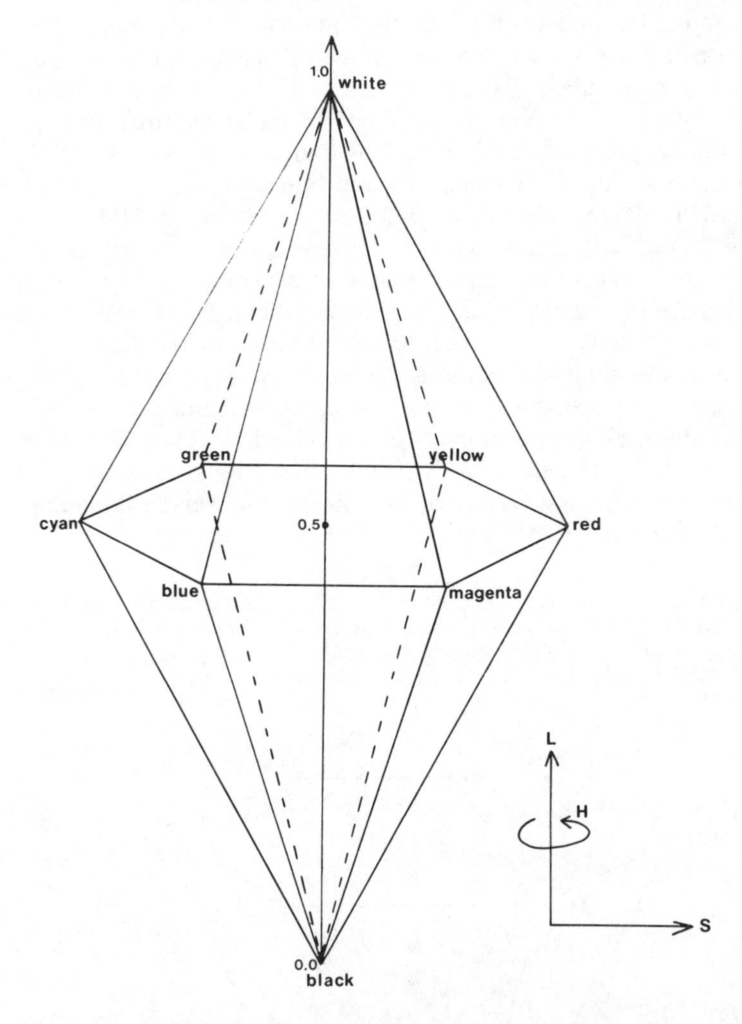

As computer hardware uses the RGB system, it is necessary to perform a conversion before displaying the appropriate colours. Algorithms for converting from HLS or HSV systems to the RGB system are to be found in Foley and Van Dam (1983). Graphics software packages using one of the colour representation systems (such as Plot 10) usually include the relevant algorithms in the code supplied with the package.

Liquid Crystal Displays (LCDs)

Liquid crystal display cells can be arranged in an array of dots, so that each dot may be turned on or off. One of the problems of LCD displays is that the display does not produce light, but reflects ambient light. Although the power requirements of such displays are thus low, it can make them difficult to read, and this has been found with the display screens on the portable computers with which LCDs have become popular.

2.4 DISPLAY PROCESSORS

As we saw at the beginning of this chapter, a display processing unit (DPU) is the section of the computer that is responsible for drawing and maintenance of the image on the display device. A DPU is not necessary with a plotter or alphanumeric printer, as streams of instructions can be directly sent to these devices from the computer via an interface.

The main task of the DPU is to provide a control mechanism to drive a CRT. DPUs for raster and random scan CRTs differ markedly, and in a general-purpose microcomputer the same 'video control' DPU will be responsible for control of all screen display operations, from text output to the generation of points and lines. In this section we will consider two types of DPU: dedicated 'graphics' random scan, and raster DPUs.

Vector or Random Scan DPU

The most basic operation of a vector DPU is to direct the electron beam to a single (x,y) point on the CRT screen. The simplest variation of this type of DPU therefore contains two registers which contain x and y values for a series of points. The voltages to move the electron beam to the given (x,y) points are sent to the CRT when the registers are read (Figure 2.10). The problem with this arrangement is that few applications need to display only a few points. More usually, primitives like lines or characters are to be displayed. A 'point plotting' DPU has to rely on CPU-controlled programs to perform the operations to interpolate points along lines, before loading these points into the x and y registers. Operations of this kind are slow, and tie up the CPU.

The obvious alternative is to allow the DPU itself to carry out interpolations and character generation. The DPU thus becomes an autonomous processor, and the additional hardware to produce the line data is called a vector generator. A pair of new registers are also added to hold the endpoints of the lines to be drawn. Vector generators work extremely quickly – up to 30 million increments per second. Characters are stored in a block of memory accessed directly by the DPU. The normal method of character storage is in the form of a matrix similar to that used in the formation of characters on a dot matrix printer.

Raster Scan DPU

Refresh DPUs are much simpler to encode because there is a direct rela-

Figure 2.10: Structure of a Simple Vector Display Processor

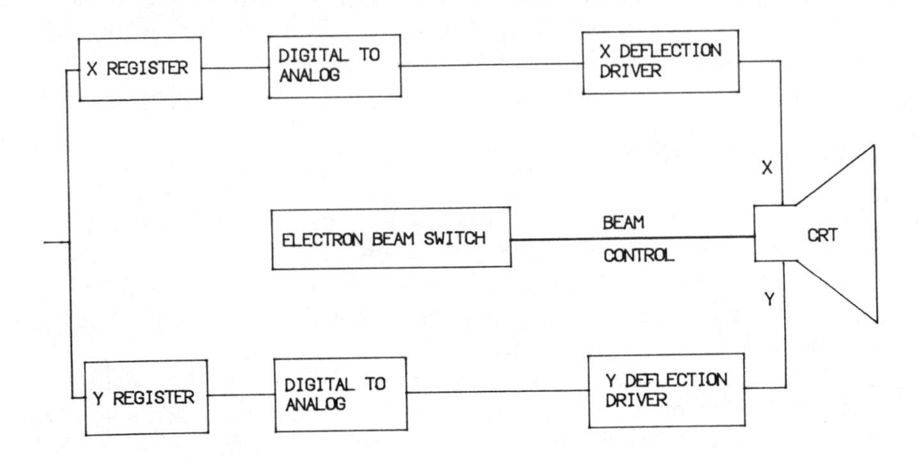

tionship between a section of refresh memory in the DPU and the arrangement of pixels on the screen. The contents of these memory locations are used to control the CRT beam's intensity. A raster scan generator produces the deflection signals which generate the raster scan: the generator also accesses the addresses in memory containing the values (that is colour value or brightness) of the pixel to be written to the screen at each scan point (Figure 2.11).

The simplest type of refresh (or frame) buffer contains one bit per pixel, representing two colours (black or white). It is now common to obtain refresh CRTs with colour capability, and apart from the expense of using a colour CRT, the main limitation to 'glorious technicolour' images is the storage needed for each pixel. The basic rule is that the number of colours available is two raised to the power of the number of bits available for each pixel; two bits therefore gives four colours, and eight bits gives 256 colours. The memory needed will double for each extra bit per pixel, and this is the reason that cheap 'home' computers rarely offer more than four or eight colours simultaneously on the screen.

The refresh DPU that we have considered so far has a major shortcoming over the random scan DPU: it has no ability to interpolate line data from endpoints or to create other graphics primitives. At some point in the creation of an image, the same kinds of algorithms that are embedded in the DPU to perform vector generation must also be done on the raster graphics data. There are several ways in which the necessary instructions can be incorporated into the system. In the simplest case (as we will see in the next chapter), a high level interpreted language instruction set can be augmented with 'graphics commands' which access machine code sub-

Figure 2.11: Structure of a Colour Refresh Display Processor

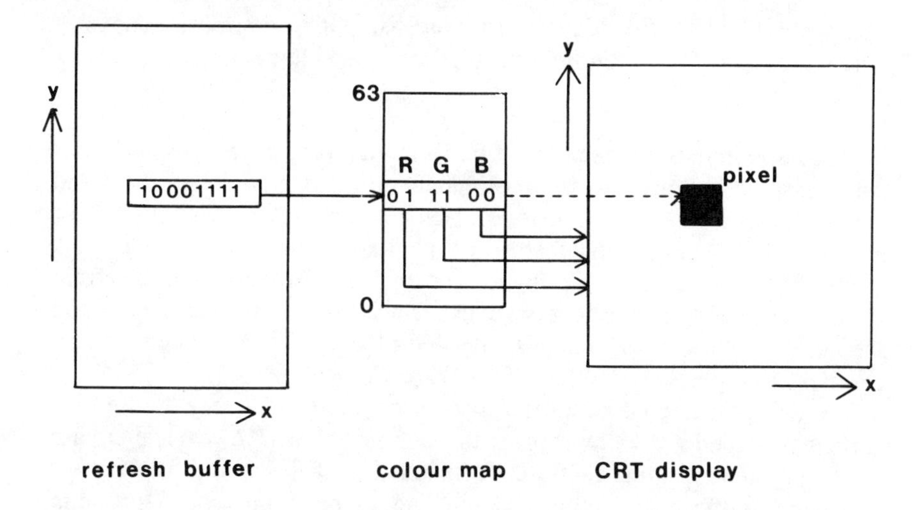

routines that perform the necessary operations during program execution. This method is used on home microcomputers. The advent of cheap raster devices based on TV technology has increased the market for random scan displays at the lower resolution (for example 512 × 512 pixel) level. Indeed, the technology has become so cheap that even home computers costing under £100 now have the capability to display resolutions of 320 × 200 pixels in four simultaneous colours.

This discussion concludes our survey of graphics DPUs. We will, however, return to this topic in the next chapter in relation to graphics software. The DPU provides the interface between computer program and display and an awareness of DPU function is important if graphics software is to be viewed from the correct perspective.

2.5 THE COMPUTER

We have now worked our way through the family of graphic devices to arrive at the computer itself. Most computers perform a variety of computational tasks and are not limited to picture processing (apart from so-called 'graphics workstations' – see below). We will divide computers therefore into two categories: minicomputers and mainframe computers, which are interfaced to graphic displays, and microcomputers which already have a DPU of the type outlined above built into their circuitry. This distinction is very blurred, however. The larger size class of computer encompasses everything from the relatively small DEC PDP-11 through to the extremely powerful CRAY machines. Graphics systems using large computers as hosts can be further subdivided into two distinct groups:

(1) those which provide local processing of graphical information;
(2) those which rely entirely on the host machine to perform graphics processing.

Microcomputers also cover a wide performance range. They usually contain a raster CRT used for both alphanumeric and graphics display, and vary from small home computers to machines costing tens of thousands of pounds. The distinguishing feature of a microcomputer is of course the single-chip processor. At the higher end of the microcomputer spectrum are the so-called 'graphics workstations', which often feature both hard and software graphics enhancements like dedicated graphics processors and libraries of graphics subroutines.

You will learn much about the sorts of computer that may be used for particular graphics tasks by reading through this book. However, there are some very general hints that are worth making at this point. The first of these is that the more 'number crunching' is performed on a program before the graphics image is produced, the more important becomes the capabilities of the computer being used. It is hopeless to attempt a simulation requiring simultaneous solution of large differential equations using a small minicomputer interfaced to a high resolution random scan CRT costing perhaps twice as much as the computer; money would be better spent on a less sophisticated graphics device and a more powerful computer. Conversely, applications like molecular drawing programs may be able to function with a small local computer accessing data via a link to a large remote mainframe. In such a case it would be advantageous to use a sophisticated graphics display perhaps with the ability to perform operations like rotations on the displayed image.

2.6 REFERENCES AND BIBLIOGRAPHY

Details of specific items of hardware can be found in product descriptions and catalogues. A fuller analysis of DPUs is given in Foley and Van Dam (1982: referenced in Chapter 1). Further reading is offered below.

Bechtolsheim, A. and Baskett, F. 'High Performance Raster Graphics for Microcomputer Systems', *Computer Graphics* 14 (3) (1980), 43–7

De'ath, S.P., Jones, T. and Butcher, B.J. 'Faster Rasters', *Systems International* (April 1984), 72–3

Coleman, N. 'Graphic Account', *Systems International* (January 1984), 38–9

Foley, J.D. and Van Dam, A. *Fundamentals of Interactive Computer Graphics* (Addison-Wesley, Reading, Massachusetts, 1982)

Goddard, D. 'Colour Electrostatics', *Systems International* (March 1984), 22–4

Harrison, N. 'Raster Colour on Paper', *Systems International* (August 1984a), 34–7

—— 'Ink Jet Advances', *Systems International* (February 1984a), 27–8
Lechner, B.J. 'Liquid Crystal Displays', in M. Fairman and J. Nievergelt (eds), *Pertinent Concepts in Computer Graphics* (University of Illinois Press, Chicago, 1969)
Negroponte, N. 'Raster Scan Approaches to Computer Graphics', *Computers and Graphics* 2 (3) (1977), 179
Newman, W.M. and Sproull, R.F. *Principles of Interactive Computer Graphics*, 2nd edn (McGraw-Hill, New York, 1979)
Oakley, D. 'Monitor Quality', *Systems International* (June 1984), 39–40
Piller, E. 'Real-time Raster Scan Unit with Improved Picture Quality', *Computer Graphics*, 14 (1) (1980), 35–8
Preiss, R.B. 'Storage CRT Display Terminals: Evolution and Trends', *Computer* 11 (11) (1978) 20–8
Sherr, S. *Electronic Displays* (John Wiley, New York, 1979)

3 Graphics Software

3.1 CONNECTING COMPUTERS AND GRAPHIC DEVICES

The usual meeting ground between computer and user is the computer operating system (we will assume that you are familiar with the procedure for entering, editing and running simple programs on at least one type of computer, be it a simple home micro or a two million pound IBM mainframe). The dialogue between user and computer takes place using text input from a keyboard: the computer prompts the user for input, and the input determines the computer's response. A variety of graphic devices may be interfaced to the computer, but unless the user is aware of the relevant commands to access and drive the graphics devices they will remain idle. There is indeed a mystique to computer graphics among the uninitiated, and often the larger the computer, the less accessible seems the possibility of using graphics. Multiaccess mainframe computer systems often provide only alphanumeric terminals for general use, and any graphic displays may be sequestered away in a little-used room or suite. In the worst case there may be little or no provision for graphics work to be done.

We have already seen that devices for graphic input and output can be inexpensive to purchase, and there is no reason why a graphics CRT could not, in principle, be linked to any mini or mainframe computer. There are two provisions to this simple statement. The first is that a method of physically connecting device and computer must be available. The connection is called a standard interface, and various protocols exist for ensuring that the right signals pass in the correct way between computer and peripheral. We will not deal further with interfacing in this book, but you will find several useful references at the end of this chapter.

You will remember from the previous chapter that graphic peripherals require instructions to either drive the drawing mechanism on an output device or to receive and process data from the sensing mechanism on an input device. Special routines are necessary to perform these tasks, and they are called (logically enough) graphics device drivers. On a microcomputer with graphics capability there is normally no need to worry about drivers, as special code will have been built into the machine's operating system. On a mainframe computer no such easy way out can be taken: although the protocols accepted by some graphic devices, most notably the Tektronix 4010, have become part standards, and you will often find that many graphics devices emulate a variety of other graphic devices. Unfortunately, the largest (and rival) graphics equipment manufacturers do not wish to provide help for the competition, so you will not find Calcomp devices emulating Tektronix terminals, and vice versa.

Let us consider the nature of device drivers in a little more detail. A device driver is a set of low-level software that handles all the device-dependent operations. Drivers are used in conjunction with other software, for example graphics libraries (see below) to make the software device independent. As an example consider drawing on a raster CRT and a plotter. At first glance, these operations are highly device-dependent, but use of a suitable device driver provides an interface between device and graphics software, so that the same software can run on both types of device.

When a program is compiled and subsequently run, the first routine that must be called in the graphics library is one that sets up the driver for the device being used. For instance if you are using the Tektronix Plot 10 Interactive Graphics Library (IGL) to drive graphics devices, the following call will instruct the terminal in use to enter graphics mode.

```
CALL IGLINI (IDEV,ID,IB)
```

where the routine called IGLINI is called using the parameters defining device type (IDEV), device option (IO) and baud rate (IB). This particular graphics library includes drivers for the various Tektronix graphic devices.

Subsequent calls will cause the various graphic operations to be performed, and we will look at some of these later in this chapter. In general we can say that calling a graphics library routine results in the following operations.

(1) Machine code instructions and data pass from computer to graphics device.

(2) The DPU (see Chapter 2) recognises the encoded graphics instructions and loads the x and y registers with any coordinate data sent with the instructions.

(3) The DPU performs the required operations and signals back to the computer that the routine is completed.

These operations are of course relevant only for graphic output. If input of data is required (from a light pen or digitizer for example), the graphics instructions sent from the computer will fetch the input data from the correct registers in the DPU.

As you will see in the next section, complete libraries of graphics routines can be expensive to purchase and update. A 'rough and ready' alternative for the machine code programmer is to write graphics drivers for a particular device from scratch. The instructions needed for creation of primitives like points and lines, and to move the graphics cursor are normally to be found in the manuals supplied with the device. This is a seemingly cheap option, but you will have to bear in mind the hidden costs of the programmer's time involved in writing the necessary drivers and routines.

By this stage the reader with a small microcomputer will probably be sighing with relief that he or she does not have to worry about graphics drivers. Unfortunately, the most complex graphics applications still have to be performed on more powerful computers; the more use you find this book, the more chance that you too will be cursing lack of device standardisation in the future.

3.2 GRAPHICS SOFTWARE PACKAGES

The first point that should be made about graphics software concerns graphics standards. For many years a kind of 'tower of Babel' has existed around provision of graphics software. Unlike the situation with device drivers, which presumably will always be different for varying items of hardware, there is no real need for graphics programmers with different graphics packages to use different instructions to, say, draw a line from A to B. There is in fact a move away from device-dependent to device-independent packages to induce portability of applications programs.

Two systems of graphics standards are at present in use to varying degrees. These are called the 'Core Graphics System' (Graphics Standards Committee, 1979), and the 'Graphics Kernal System' (GKS) — see Encarnacao *et al.* (1980), International Standards Organization (1981), Spiers (1984). The GKS system can in fact be viewed as a 2D subset of the Core software. Computer manufacturers are coming under increasing pressure to adopt a standard. In future, therefore, it may be possible to write a program in a high-level computer language like Fortran, and to make calls from this language to a graphics library secure in the knowledge that the program will run using any other graphics library conforming to the given standard. Although there are differences between Core and GKS we will not discuss them here, and it is likely that one of the two systems will gain supremacy in the next two or three years.

Our survey of graphics packages will begin with two types of graphics package available for mainframe computers, namely graphing packages offering limited 'graph drawing' facilities, and full-blown graphics libraries allowing a variety of 2D and 3D manipulations of data. After this, we consider graphics available on general purpose microcomputers. The final class of graphics software is that found on dedicated 'graphics workstations', perhaps representing the graphics' 'state of the art'.

3.3 GRAPHICS PACKAGES ON MINICOMPUTERS AND MAINFRAME COMPUTERS

The most elementary type of package includes a basic set of commands for creation of graphs on an output device. Such packages are often among the most 'user friendly' items of graphics software and may contain routines to simplify the drawing and labelling of graph axes and other graphic aids of use in the analysis of scientific data. The specialised nature of graphing as opposed to more general graphics packages precludes them from joining any graphics standard system. We will look at two such packages here. The first, called PLOTALL, caters for the less computer-literate user, and is available from the University of Akron, Ohio (Seymour and Wiggins, 1981). The PLOTALL language was developed to allow the user to communicate graphical instructions in 'English-like' statements, and therefore obviates the need to know any computer language at all. It is possible to produce two dimensional graphs, histograms, scatter diagrams and pie charts using PLOTALL, and a PLOTALL program consists of four distinct sections as follows:

(1) data description;
(2) data;
(3) plot description;
(4) the PLOT command.

Here is a simple PLOTALL program:

```
1 { THE VARIABLE IS TEMP
  { THERE ARE 5 OBSERVATIONS
  { READ THE DATA
  { 30
2 { 40
  { 50
  { 45
  { 60
3 { THE PLOT TITLE IS 'TEMPERATURE'
  { THE TYPE OF PLOT IS LINE PLOT
4   PLOT TEMP
```

Note that section 1 sets the name of the data set ('TEMP'), together with the number of observations to be read. Section 2 contains the successive values of the variable 'TEMP' prefaced by an instruction to read the data. Section 3 contains a description of the plot type and title, and section 4 instructs the computer to begin processing the data.

There are three restrictions concerning the placement of the PLOTALL commands. These are:

(1) The data description must precede the data.
(2) The PLOT command must follow the plot description statements.
(3) The data must appear sequentially without plot description statements breaking them up.

At first glance, PLOTALL appears too be very sophisticated: after all, how many other computer languages enable you to input sentences in English? This sophisication is in fact illusory, as the breakdown of the following statement shows.

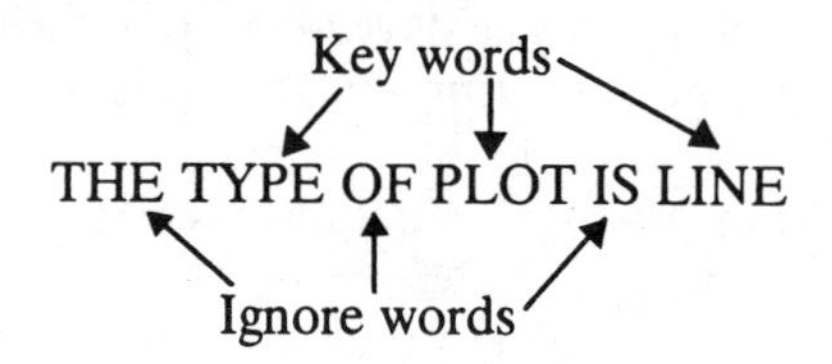

You could therefore write this statement as,

TYPE PLOT LINE

with more economy. The value of the 'English' in PLOTALL is therefore a matter of taste.

PLOTALL allows a fair degree of flexibility with regard to plotting. Besides specification of the various plot types, it is also possible to specify a number of different options. These options include size, location and rotation of text, title and plot; plotting of different symbols; shading patterns for histograms and pie charts, and choice of curve fitting algorithm (linear, quadratic, cubic, quartic or spline). Here are two PLOTALL sample programs (programs 3.1 and 3.2; and Figures 3.1 and 3.2) with a biological flavour. Note how readable the programs are!

A rather more flexible route to graph preparation is offered by SIMPLEPLOT (Bradford University Software Services Ltd, 16 Campus Rd, Bradford BD7 1HR, UK), which is widely used in scientific work. Unlike PLOTALL, SIMPLEPLOT is a library of graphics subroutines calleable from Fortran and Pascal programs, and therefore sits somewhere between being a graphing package and a full graphics library. This package started life in the early 1980s as an elementary collection of routines to simplify graph preparation, but has now grown into a library of some 250 subroutines. The SIMPLEPLOT library is split into five separate sections as follows.

Program 3.1

```
THE VARIABLES ARE WILDTYPE,XW,S.OCULIS,XS,EYE-2,X2,EYE-D,XD
THE NUMBER OF CASES IS 6
READ THE DATA
 8,72, 6,72,16,72, 4,72
14,84, 8,84,23,84,10,84
17,96,34,96,21,96, 9,96
14,108,95,108,40,108, 2,108
13,120,30,120,49,120, 3,120
22,132,28,132,35,132, 4,132
TITLE 1 IS 'NUMBER OF DEAD CELLS SEEN'
TITLE 2 IS 'IN HISTOLOGICAL SECTIONS'
TITLE 3 IS '  OF IMAGINAL DISKS IN'
TITLE 4 IS '  DROSOPHILA GENOTYPES'
THE X-AXIS SCALE IS 72,132
THE Y-AXIS SCALE IS 0,100
THE X-AXIS LABEL IS 'AGE AT FIXATION (H)'
THE Y-AXIS LABEL IS 'NO. OF DEAD CELLS'
THE LEGEND LOCATION IS 6.5,6
PLOT WILDTYPE, S.OCULIS, EYE-2, EYE-D VS XD
STOP
```

Program 3.2

```
THE VARIABLES ARE AGE,CLONESIZE
THERE ARE 9 CASES
READ THE DATA
12,44
24,30
36,18
48,10
60, 8
72, 4
84, 0.7
96, 0.6
108,0.2
THE X-AXIS LABEL IS 'AGE AT IRRADIATION (H)'
THE Y-AXIS LABEL IS 'CLONE SIZE (OMMATIDIA)'
THE X-AXIS SCALE IS 0,120
THE Y-AXIS SCALE IS 0.1,100.0
Y-AXIS TYPE IS LOG
TITLE 1 IS 'FINAL SIZE OF CLONES MARKED'
TITLE 2 IS '     BY X-RAYS DURING'
TITLE 3 IS '  DROSOPHILA DEVELOPMENT'
THE LEGEND LOCATION IS 4,4
PLOT CLONESIZE VS AGE
STOP
```

Figure 3.1: A Graph Generated Using the PLOTALL Program 3.1

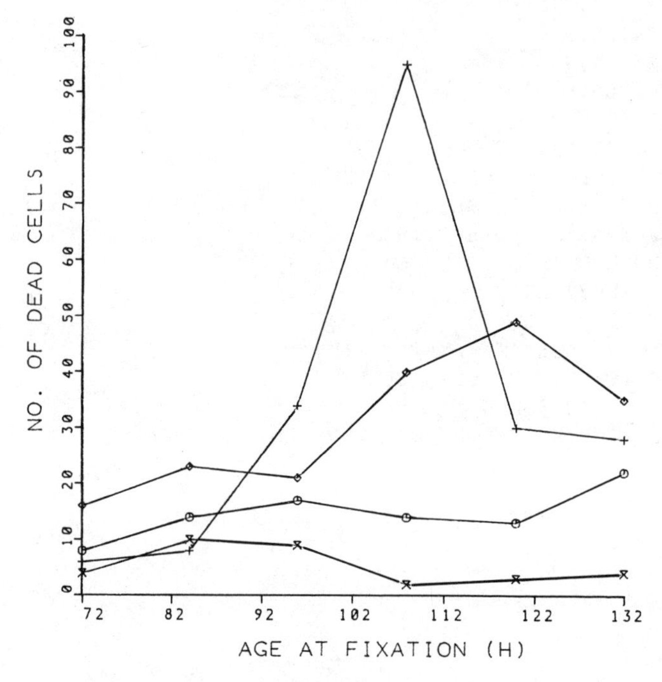

(1) Plotting two-dimensional data.
(2) Additional subroutines for plotting surface data.
(3) Additional subroutines for plotting COMPLEX data.
(4) Presentation graphics.
(5) Plotting of three-dimensional real arrays.

The SIMPLEPLOT subroutines fit into the following classification based on the type of operation.

Labelling Subroutines. These add labels to the graphs, and may be invoked at any time during graph drawing.

Plotting Subroutines. These perform the actual drawing operations, for example:

```
JOIN PT(X,Y)
```

draws a straight line from the current pen position to the point X,Y.

Figure 3.2: A Graph Generated Using the PLOTALL Program 3.2

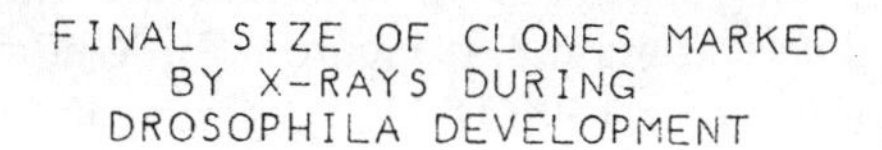

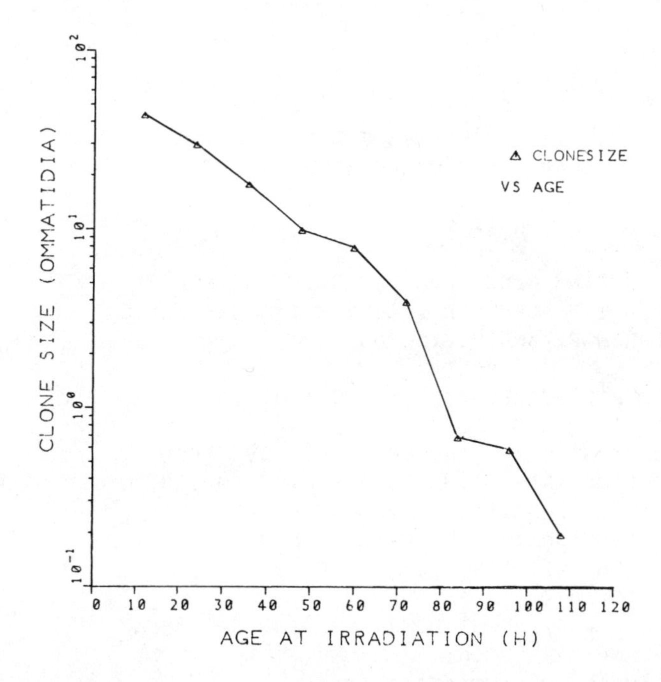

Change of Coordinate System. These subroutines allow cartesian, degrees and radians to be specified.

Reset Drawing. The reset subroutines modify plotting details, for example to change subdivision of the axes or to set different pen pointers.

Reset Scale. Subroutines in this category allow changes in X and Y axis scales.

Reset Layout. These subroutines enable graphs to be plotted singly or in groups, and to change the sizes of the graphs.

New Picture Subroutines. This category handles the drawing of the graph axes (subroutine AXES), or alternatively subroutines can be invoked to draw a polar framework or to start a picture without axis creation.

Auxiliary Subroutines. These subroutines include facilities for calculating polynomials or frequency tables for plotting histograms.

Rather than give a full listing of all the SIMPLEPLOT routines here, we give the following example to demonstrate its use in plotting two-dimensional data (program 3.3; Figure 3.3). You may like to compare this program with the second PLOTALL program above.

Program 3.3

```
C       THIS IS A SIMPLEPLOT DEMONSTRATION PROGRAM
C       TO DRAW A LOG GRAPH WITH RANGES
C
        DIMENSION AX(9),AY(9)
C
C       DEFINE LINEAR VS LOG PLOTTING SCALES
C       SET SCALE TYPE 1=LINEAR,2=LOGARITHMIC,
C       RANGES AND DIVISIONS
C
        CALL SCALES (0,0,108.0,1,0.1,50.0,2)
        CALL XAXDIV (12.0)
C       START GRAPH AND DRAW PAIR OF AXES
        CALL AXES ('AGE AT IRRADIATION',18,'CLONE SIZE',10)
C       READ 9 SETS OF 2 VALUES
        OPEN(UNIT=21,FILE='SIMPLE.DAT')
        READ (21,1000) (AX(I),AY(I),I=1,9)
        DO 10 I=1,9
C       DRAW STRAIGHT LINE TO POINT JUST READ
C       AND MARK POINT
        CALL JOIN PT(AX(I),AY(I))
        CALL MARK PT(AX(I),AY(I),1)
10      CONTINUE
C       RAISE PEN AFTER LAST LINE IS DRAWN
        CALL BREAK
C       NOW PLOT RANGE FOR EACH POINT
        DO 20 I=1,9
        READ (21,1000) YMIN,YMAX
        CALL RANGE (AX(I),YMIN,AX(I),YMAX)
20      CONTINUE
C       WRITE A TITLE AT TOP CENTRE OF GRAPH
        CALL TITLE('H','C','FINAL SIZE OF CLONES MARKED',27)
        CALL TITLE('H','C','BY X-RAYS DURING',16)
        CALL TITLE('H','C','DROSOPHILA DEVELOPMENT',22)
C       TERMINATE PLOTTING
        CALL END PLT
        CLOSE(21)
        STOP
1000    FORMAT (2F10.4)
        END
```

Figure 3.3: A Graph Generated Using SIMPLEPLOT Routines Called from a Fortran Program (Program 3.3)

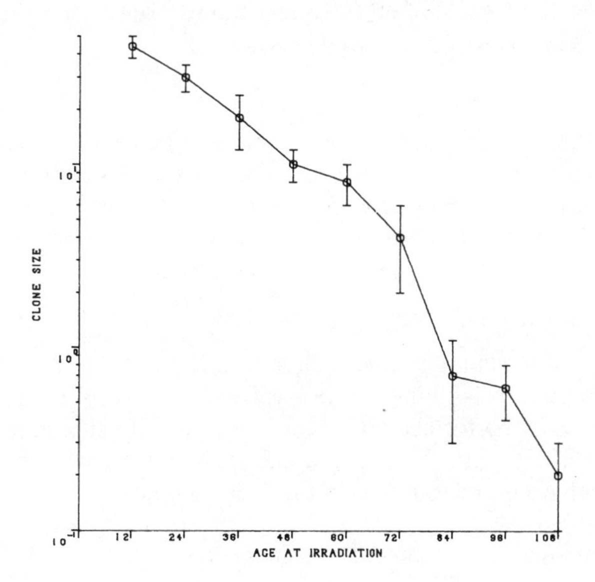

The Tektronix Plot 10 IGL library (Tektronix Inc, Beaverton, Oregon, USA) represents the next, more flexible level of graphics package. Unlike SIMPLEPLOT, IGL is designed to be used by the graphics professional, and includes a number of sophisticated routines for manipulation and storage of graphics images. The basic features of IGL can be used to draw elementary shapes, but the full library also includes routines for transformations in two and three-dimensional space, for manipulating segments, and to draw text in a variety of different founts. You might like to note, however, that for all its sophistication, IGL will not automatically draw graphs for you!

As with SIMPLEPLOT, IGL subroutines are most commonly called from Fortran although it is possible to use Pascal for the main program. IGL routines are classified according to function, and the main groups are as follows.

(1) System environmentals.
(2) Graphics environmentals.
(3) Graphic action.
(4) Text environmentals.
(5) Text action.
(6) Host file communications.
(7) Utilities.

A restricted set of routines handle two-dimensional graphics for a single text font. Options are also available for three-dimensional graphics, segment display and for use of multiple text founts. The important features of the various routine categories are discussed below.

System Environmentals

Various IGL environmental conditions have to be set up before or during program execution. The graphics terminal has to enter or exit graphics mode, and the correct device code and baud rate must be chosen. Other system environmentals allow IGL output to be transformed to or from permanent files and to temporarily exit and re-enter IGL.

Graphics Environmentals

Graphics environmental routines directly affect the form of the graphic output in some way. Routines of this type are 'passive' in the sense that no graphic action is performed, but the form that any subsequent graphics action takes is determined by the graphics environmental routines invoked. The main subgroups of routines do the following tasks.

For Two-dimensional and Three-dimensional Operations.

(1) Units (graphics display and angles).
(2) Drawing characteristics (dashed or continuous lines, selective line erasure, line and background colour).
(3) Clipping.
(4) Transform environment (world space or viewing surface only).
(5) Modelling transforms (scale, rotation, translation, matrix transformations).
(6) Line smoothing.
(7) Panel filling.

For Three-dimensional Operations Only.

(1) Billboard creation.
(2) Projections (parallel or perspective).
(3) Eye position.
(4) Modelling transforms (three-dimensional equivalents to two-dimensional transforms plus shearing, etc.).

Graphics Action

The central IGL routines are those which display the image. Routines are included to perform the following operations.

(1) Graphics output — that is, moving the graphics cursor, drawing a

line from the current cursor position to a specified point, drawing of polygons, arcs, panels and markers.

(2) Graphics input – location of a digitized point.

(3) Other routines – screen erasure, return graphics cursor to origin, ring terminal bell, generate hard copy of screen contents.

Text Environmentals

Although IGL text can be displayed on screen or plotter at the same time as graphics output, the graphics environmental routines do not as a rule affect text. Conversely, text environmental routines do not affect graphics output. The main groups of text environmental routines are concerned with the following tasks.

(1) Text appearance (size, fount, quality, colour).
(2) Character spacing.
(3) Fount specification.
(4) Text orientation.
(5) Cursor positioning.
(6) Coordinate system (is text to be affected by modelling transforms?)

Text Action

Text input and output are accomplished with these routines.

Host File Communications

These routines are concerned with opening, closing and checking communications channels, reading files and writing to files.

Utilities

The power of IGL is enhanced by the inclusion of a number of utility routines which can perform transformations of coordinates between coordinate systems, or may convert data into different forms of representation (such as integer to character string). The general classification of these routines is as follows.

(1) Data conversion utilities.
(2) Character conversion utilities.
(3) Fount manipulation.
(4) String length calculations.
(5) Line smooth emulation.

The full Plot 10 Interactive Graphics Library totals some 330 routines. It is therefore a comprehensive list of graphics operations. Even so, there are some limitations, notably an inability to handle hidden line and surface manipulations, although these may soon be available.

Two examples of the use of Plot 10 IGL in a biological context may be found in Chapter 7, programs 7.2 and 7.3.

3.4 MICROCOMPUTER GRAPHICS SOFTWARE

As we saw in the last chapter, the relationship between graphics processor and host computer may vary. The packages that we have so far considered run on mainframe or minicomputers coupled to display terminals that act as 'dumb' devices only capable of drawing the display information sent from the main computer. The second type of relationship between host and graphic device is that seen in microcomputers, where CPU and DPU are both present in the same box. This arrangement has two advantages. Firstly, the graphics device driver is 'built in' to the machine, and secondly the graphics software may be specifically tailored for the capabilities of the graphics hardware.

'Personal' computers like the Commodore 64, Apple II or IBM PC have graphics commands embedded into ROM that can be directly accessed as keywords by the Basic interpreter. If you forget the unpleasantness of writing more than a 50-line program in Basic, personal computer graphics present an excellent way of learning how to use the fundamental graphics technique to be found in Chapters 4 and 5 of this book. It is very convenient to be able to write ten or so lines in Basic and to be immediately rewarded by a graphic image generated just by typing 'RUN'. In essence, however, there is not much real difference in programming in this way than in using the simpler routines in a package like IGL. Compare the following programs to draw a triangle defined by coordinates (X1,Y1), (X2,Y2), (X3,Y3) in Commodore Basic V7.0 (Commodore 128) and Fortran calling Tektronix IGL routines.

First, the Commodore version ...

```
 5  REM PROGRAM TRIANGLE
10  GRAPHIC1,1
20  LINE X1,Y1,X2,Y2,1
30  LINE X2,Y2,X3,Y3,1
40  LINE X3,Y3,X1,Y1,1
50  STOP
```

And the IGL program ...

```
PROGRAM TRIANGLE
CALL IGLINI(4010,1,9600)
CALL NEWPAG
CALL MOVE(X1,Y1)
```

```
CALL DRAW(X2,Y2)
CALL DRAW(X3,Y3)
CALL DRAW(X1,Y1)
STOP
END
```

A typical range of graphics commands on microcomputers may be found on the IBM Personal Computer. On this machine a colour/graphics adaptor is needed to use graphics: two 'graphics' modes are possible, 'medium resolution' (320 × 200 points in up to four colours), and 'high resolution' (640 × 200 points in two colours). The following Microsoft Basic commands are available in the graphics modes.

*	SCREEN	sets up a graphics screen
	DRAW	draws a shape
*	LINE	draws a line
*	COLOR	sets colour of line/background
	POINT	returns colour of a given point
	GET	inserts colour of points in a defined region into an array
	PAINT	paints an area with colour
*	PSET	plots a point
	PUT	writes data stored with GET onto the screen
*	CIRCLE	draws a circle or ellipse

Commands to allow input from a light pen and joystick are also available. The functions of the asterisked commands listed above are given below.

```
SCREEN [mode],[burst],[apage],[vpage]
```

Mode sets the graphic mode as 0 (text), 1 (medium resolution) or 2 (high resolution). Burst enables colour (1) or disables colour (2). Apage (active page) and vpage (visual page) select pages to be displayed (ie areas of memory to be used as data for the bitmap). Any of these parameters may be omitted, for example

```
SCREEN 0,1
```

selects text mode with colour, while

```
SCREEN 2,,0,0
```

switches to high resolution graphics mode and sets active and visual pages to 0.

LINE [x1,y1]–(x2,y2) [,[colour] [,B[F]]]

draws a line from (x1,y1) to (x2,y2). Note that the only mandatory parameter is a single pair of coordinates, so that

LINE –(x2,y2)

draws a line from the last referenced point to the point (x2,y2).

The colour parameter in the LINE statement is an integer from 0 to 3, and selects colour from the current colour palette as defined by the COLOR statement (see below). 'B' allows the user to draw a rectangle with the points (x1,y1) and (x2,y2) as opposite corners of the rectangle. 'BF' draws a rectangle filled with the selected colour.

COLOR [background] [,[palette]]

Background specifies the background colour chosen from the available colours 0 to 15. The choice of foreground colours in graphics mode is limited to those available in two preset palettes, 0 and 1:

Colour	Palette 0	Palette 1
1	Green	Cyan
2	Red	Magenta
3	Brown	White

Note that COLOR may only be used in graphics mode 1 (medium resolution). An error results if you try to use it in high resolution mode.

CIRCLE (x,y),r[colour[,start,end[,aspect]]]

This command draws a circle around the point (x,y) with radius r. Arcs may be drawn by specifying the start and end angles in radians. Aspect affects the ratio of the x radius to the y radius, allowing ellipses to be drawn.

PSET (x,y)[,colour]

Plots a point at location (x,y). Colour may be selected in medium resolution mode only.

This simple set of graphics commands is far short of the complexity inherent in mainframe graphics libraries, but provides the basis for some quite advanced graphic effects to be obtained. The limitation of this type of microcomputer lies less in the graphics commands available, but more in the low power and limited graphics resolution.

More expensive and powerful microcomputers that overcme these shortcomings often have their own graphics libraries that must be linked into programs before execution. The DEC Professional range for example uses a 'Core Graphics Library' for this purpose, and the routines are invoked by calls similar to those in Tektronix IGL.

3.5 GRAPHICS WORKSTATIONS

The most advanced microcomputer graphics performance is offered by 'supermicro'-based graphics workstations. These machines are often closer to superminicomputers in performance, for example the DEC MicroVax II which contains the VAX 11/780 processor 'shrunk down' to a single chip. More conventional microprocessor-based machines use advanced 32-bit chips like the Motorola 68000 series. The high performance microprocessor is usually linked to a raster display via a frame buffer. We will look here at the graphics software offered on one representative range of graphics workstation: the Apollo Domain (Apollo Computer Inc, Chelmsford, Massachusetts, USA).

As you might expect from a range of machines aimed at the graphics market, the graphics software is comprehensive and three package options are available on all Apollo machines. These are:

(1) Graphics Metafile Resource (GMR)
(2) Graphics Primitive Resource (GPR)
(3) Domain Core Graphics.

GMR allows the user to create device-independent files of picture data. These files can be edited, displayed and stored, and a full range of transformations and other operations can be rapidly performed on the filed data. The coordinates used are device-independent, giving flexibility in program execution. Each file of picture data is divided into segments, each segment consisting of a sequence of commands. A metafile is defined as a self-sufficient low-level but still device-independent picture description (Foley and Van Dam, 1982). Metafiles are therefore very useful if picture descriptions are to be transported between devices or installations. The outline structure of a GMR program looks like this:

(1) Initialise GMR.
(2) Create a file.
(3) Create a segment within the file.
(4) Setup commands within the segment (to draw an object for example).
(5) Perform segment commands to draw picture.

(6) Close the metafile.
(7) Terminate GMR.

GPR differs from GMR mainly in that changes are made directly to the bitmap rather than to a metafile system. There is thus no memory of the calls performed unless a static image is stored at a given time, and re-drawing requires that the application program itself keeps track of the calls made. The display coordinates are device-dependent. GPR is roughly analogous to the Tektronix Plot 10 IGL library discussed earlier. The graphics primitives handle the following operations.

(1) Drawing lines.
(2) Acquiring text founts and manipulating text.
(3) Manipulating graphics with bit-block transfers.
(4) Filling polygon areas.
(5) Handling input operations.
(6) Setting attributes.
(7) Enabling direct graphics operations.
(8) Imaging with an extended colour range.

Core Graphics conform to the industry standard Core discussed earlier. Core Graphics is similar to GPR in structure, but the coordinates are device-independent to give flexibility in the development and use of application programs.

The relationship of the three graphics packages are shown in Figures 3.4, 3.5 and 3.6. A hierarchy can be seen, with the applications program sitting directly on top of GMR, while Core and GPR are at successively lower levels.

Figure 3.4: Apollo Domain Graphics Primitive Resource: Relationship to the Graphics System

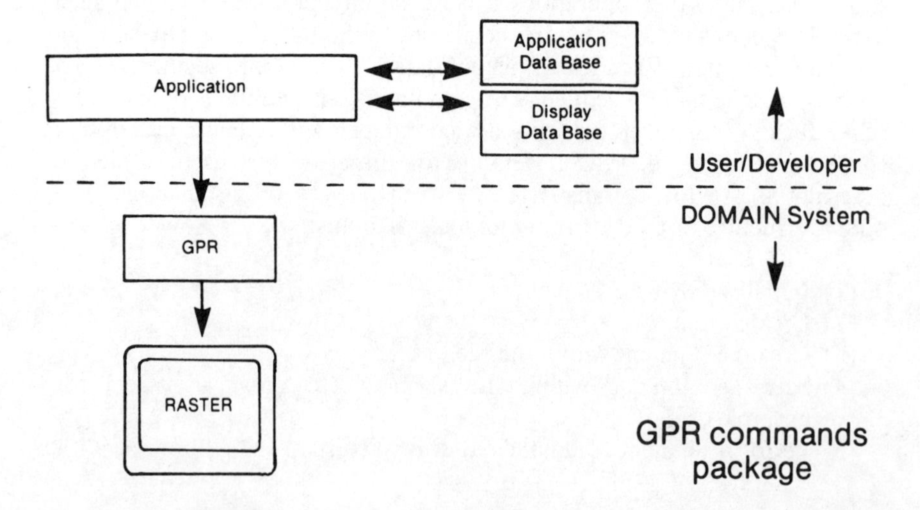

Figure 3.5: Apollo Domain Graphics Core Package: Relationship to the Graphics System

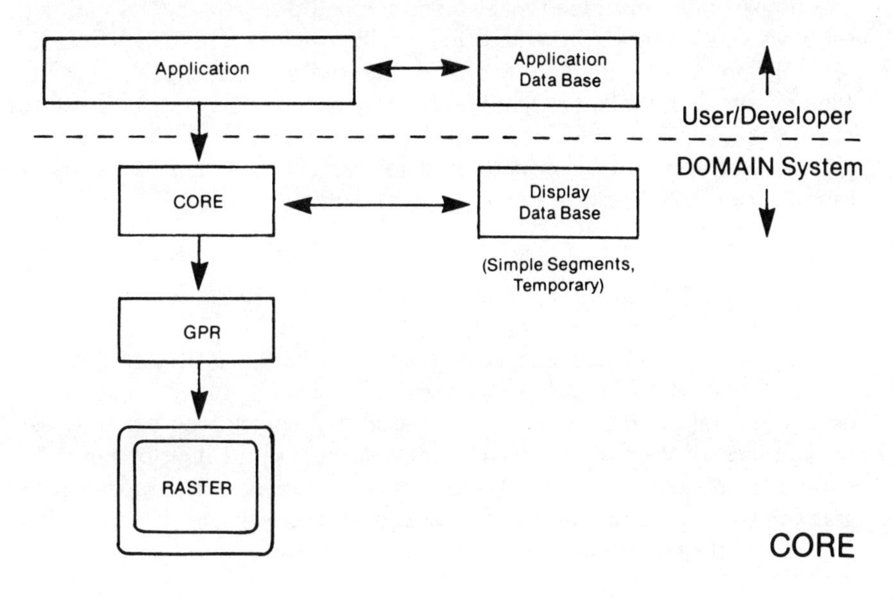

Figure 3.6: Apollo Domain Graphics Metafile Resource: Relationship to the Graphics System

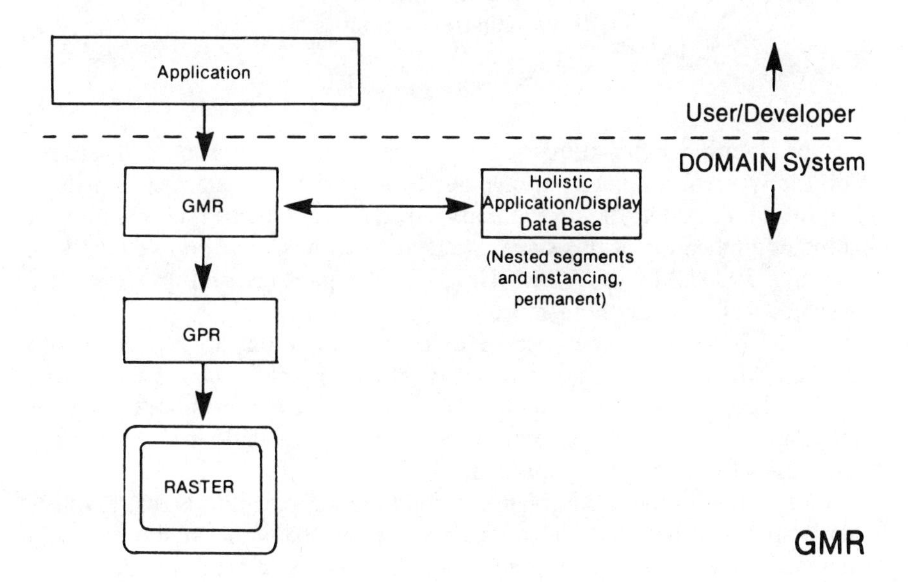

The Domain operating system supports 16 or 24 concurrent processes, depending on the machine type. The bitmapped display supports multiple windows which can be presented side by side or can be partially overlaid. Multiple processes can therefore be monitored simultaneously, and any window can be instantly brought into full screen view. The system therefore offers both sophisticated graphics hardware together with integrated graphics software, and is widely used in intensive graphics-based applications like CAD (computer-aided-design) work.

3.6 THE APPLICATIONS PROGRAM

Given a set of graphics routines, the computer user must create the particular applications program to perform a given task. In this chapter we have looked at the raw materials to be used in graphics programming, but the graphics software is very much a single element in the development of a particular application. If we view the process of interactive graphics programming as a hierarchical process, we can see the relationship of graphics software to the other components.

Graphics display
↑
Graphics software library
↑
High-level language program
with graphic transformations
↑
Applications model

Our attention so far in this book has been focused on graphics hardware and the types of graphics package available to the programmer. It may be helpful to consider the preliminary strategy to be used in planning a graphics application at this stage. Some applications, for example involving use of a PLOTALL or SIMPLEPLOT type package, require few extra graphics skills. If anything more complex than a graph is to be plotted, however, there are a number of extra considerations to be taken into account when choosing software. Can all the graphics manipulations be performed from within the graphics software package available, or does the application require spatial transformations or hidden lines/surface algorithms to be separately implemented?

One of the difficulties in planning a graphics application is that graphics hardware, software and the computer being used must all be correctly matched. It is no use attempting a graphics program that cannot be displayed on the terminal available, and the software itself must (a) be able to drive the hardware, and (b) be suitable for implementation of the relevant graphics algorithms. The computer is also very important. Some graphics applications are notoriously heavy users of CPU time, and this must be taken into account.

Graphics programming therefore requires even more forethought than general programming operations. We have previously warned of the dangers of 'jumping into the code' (Ransom and Matela, 1985). It may be necessary to plan graphics applications some months in advance of the start of a given project, in order to decide on the graphics facilities to be employed.

An appreciation of the possibilities offered by computer graphics relies to a large extent on some knowledge of geometrical manipulations. Our next task, therefore, is to consider the nuts and bolts of graphics manipulations: two and three-dimensional transformations. These topics are covered in Chapters 4 and 5 below.

3.7 REFERENCES AND BIBLIOGRAPHY

Bergeron, R.D., Bono, P. and Foley, J.D. 'Graphics Programming Using the Core System', *Computing Surveys*, 10 (4) (1978), 389–443

Butland, J. *SIMPLEPLOT Mk II Section 1: Plotting 2D Data* (Bradford University Research Ltd, Bradford UK, 1982)

Encarnacao, J. *et al.* 'The Workstation Concept of GKS and the Resulting Conceptual Differences to the GSPC Core System', *Computer Graphics* 14 (3) (1980), 226–30

Foley, J.D. and Van Dam, A. *Fundamentals of Interactive Computer Graphics* (Addison-Wesley, Reading, Massachusetts, 1982)

IBM Basic Manual, 2nd edn version 1.10 (IBM (UK), Portsmouth, 1983)

International Standards Organization *Graphical Kernal System* (GKS), version 6.6 (1981)

Michener, J.C. and Foley, J.D. 'Some Major Issues in the Design of the Core Graphics System', *Computing Surveys,* 10 (4) (1978), 445–64

— and Van Dam, A. 'A Functional Overview of the Core System with Glossary', *Computing Surveys,* 10 (4) (1970), 381–8

Programmer's Guide to Domain Graphics Primitives (Apollo Computer Inc., Chelmsford, Massachusetts, USA, 1983)

Programmer's Guide to Domain Graphics Metafile Resource (Apollo Computer Inc., Chelmsford, Massachusetts, USA, 1984)

Ransom, R. and Matela, R. *Computers in Biology: An Introduction* (Open University Press, Milton Keynes, UK, 1985)

Spiers, R. 'Using GKS', *Systems International* (July 1984), 24–5

Seymour, G.A. and Wiggins, R.A. *PLOTALL Computer Graphics Language User's Guide* (University of Akron, Ohio, USA, 1981)

'Status Report of the Graphics Standards Committee' *Computer Graphics,* 13(3) (1979)

Tektronix Plot-10 Interactive Graphics Library User's Manual (Tektronix Inc., Beaverton, Oregon, 1984)

4 Two-dimensional Graphics

4.1 THE ELEMENTS OF TWO-DIMENSIONAL TRANSFORMATIONS

Computer graphics is essentially involved with the representation and transformation of geometric objects. This might seem obvious, but a great deal of intellectual activity has gone into the investigation of forms of representation and methods of transformation. In this chapter we will discuss some of the more fundamental concepts involved in two-dimensional spatial manipulations.

The first thing that we must consider can be termed the graphic environment. In general we have an object or set of objects positioned in a predefined world coordinate system. Our ultimate goal, however, is to display all or part of this set of objects on some suitable graphics device. To accomplish this we must perform several operations, which will probably involve some form of transformation. Prior to defining the set of transformations it is necessary to define the various elements that comprise the graphics environment.

The first element that we consider is known as a window. In essence a window is nothing more than a two-dimensional rectangle enclosing the set of objects to be displayed, which are defined in world coordinates (Figure 4.1). An important aspect of the window is the ability to move it about in the world space. The implications of this are that one can selectively choose which object, objects or parts of objects you wish to observe. It is important to realise that the window acts as a boundary in the sense that the visual information in the whole scene is partitioned into two sets. The retained information is contained within the window while, for display purposes, that information outside the window is discarded.

Closely associated with the window is the viewport (Figure 4.1). The viewport consists of another two-dimensional rectangle but is defined in terms of the display or screen coordinates. By default the viewport is usually the whole screen. However, as with the window the viewport can be moved around the display surface, thus allowing for multiple images or compression or expansions of a single image.

The window and the viewport are related in the following sense. The image defined within the window is mapped or projected into the viewport on the screen. The distinction lies in the fact that the window is defined in terms of the object space, and the viewport in terms of the display surface.

The next item that we will consider, while not truly a graphic environmental, is closely related to the concept of windows and as such is presented here. The concept is that of clipping and involves the 'cutting off' of

Figure 4.1: Relationship of the 2D Graphics World, Window, Viewport and Display Surface

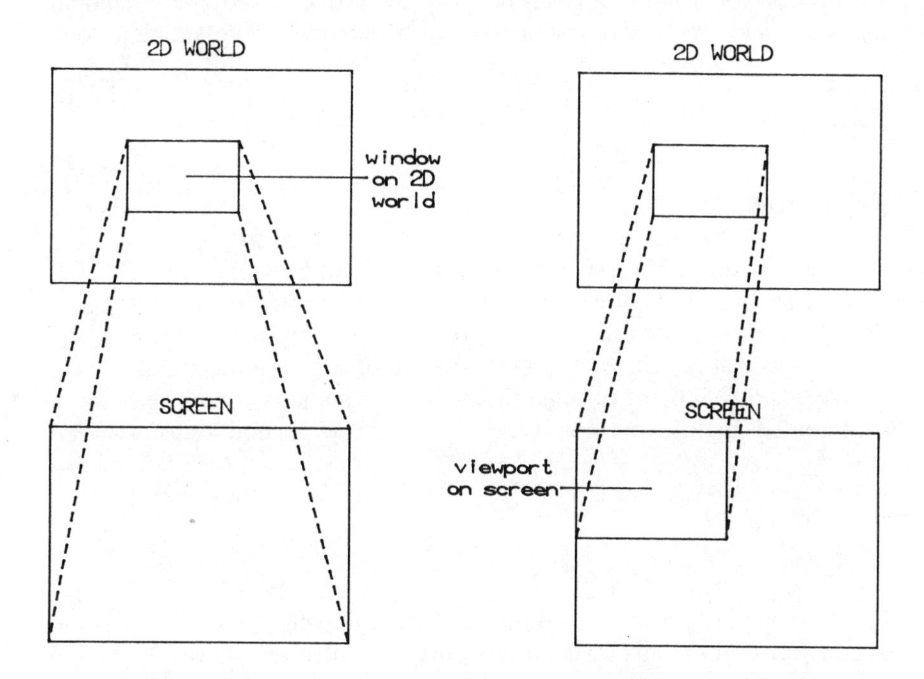

all lines that are deemed to be invisible (that is, outside the viewport). An example of implied clipping occurs within the context of windows as all line segments or other graphical information not contained within the window are made 'invisible'. There are well known clipping algorithms for handling various types of data, such as lines and polygons (Newman and Sproull, 1979). Indeed, on some advanced and quite expensive systems, the clipping can be done by the hardware itself.

We now consider some geometric quantities involved in computer graphics. For our purposes we may assume that there are basically three elementary components or primitives, from which an object will be defined. These are: points, lines and planes.

As we have previously stated, we must be able to transform these basic components. Some of the operations that might be considered would be scaling, rotation, translation, shear or perspective.

4.2 REPRESENTATION OF POINTS

A point in two-dimensional space can be represented by its x and y coordinates. The mathematical object that we use to maintain and manipulate this point information is called a matrix. In this case a one row by two column matrix is denoted by $[x \;\; y]$. The extension to three dimensions (see Chapter 5) for a point is given by $[x \;\; y \;\; z]$, a one row by three column matrix. You may also come across an alternative representation: one column by two rows $\begin{bmatrix} x \\ y \end{bmatrix}$, and one column by three rows $\begin{bmatrix} x \\ y \\ z \end{bmatrix}$, respectively.

These row and column matrices are frequently termed vectors. Assume that we have an orthogonal coordinate system and a set of points: $P1$, $P2$, $P3$, $P4$ as given in Figure 4.2. These points can be stored within a matrix and transformed or manipulated by various operations defined for matrices. See appendix A for an explanation of basic matrix operations.

We begin by looking at some simple transformations of a single point as represented by the position vector $[x \;\; y]$. The general transformation matrix is given by

$$\begin{bmatrix} a_{11} & a_{12} \\ a_{21} & a_{22} \end{bmatrix}$$

The new position vector is produced by multiplying the existing position vector (for every point) by the transformation matrix as follows:

Figure 4.2: The Orthogonal Coordinate System with Points in 2D Space

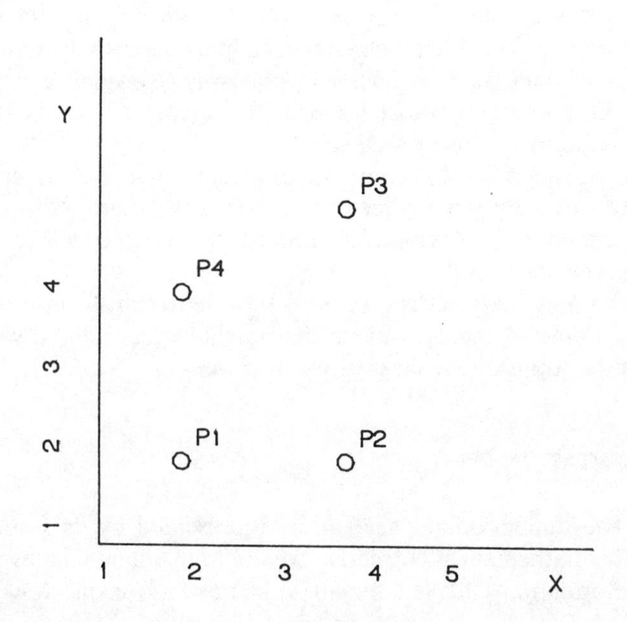

$$[x' \; y'] = [x \; y] \begin{bmatrix} a_{11} & a_{12} \\ a_{21} & a_{22} \end{bmatrix} = [(a_{11}\,x + a_{21}\,y)\;(a_{12}\,x + a_{22}\,y)]$$

where $[x' \; y']$ is the general representation of the new point after the transformation has been applied to the original point $[x \; y]$. Clearly:

$$x' = a_{11}\,x + a_{21}\,y$$
$$y' = a_{12}\,x + a_{22}\,y$$

Remember the a_{ij}s are elements of the transformation matrix where *i* is the row index and *j* is the column index.

There is a special matrix known as the identity matrix, which when applied leaves the point unchanged (invariant). For the case under consideration it would look like this:

$$\mathrm{I} = \begin{bmatrix} 1 & 0 \\ 0 & 1 \end{bmatrix} \text{ with } a_{11} = a_{22} = 1$$

and all other elements equal to zero. The a_{ii} elements are known as the main diagonal. To confirm this fact about the invariance of the identity matrix we will give an example:

$$[x' \; y'] = [x \; y] \begin{bmatrix} 1 & 0 \\ 0 & 1 \end{bmatrix}$$
$$= [(x + 0) \cdot (0 + y)]$$
$$= [x \; y]$$

If you do not understand this see Appendix 1, p. 196.

Let us now look at what happens to our point when we put some values into the transformation matrix. We first consider scaling or stretching in the *x* direction. Let $a_{11} = a$ and $a_{22} = 1$, and all other terms equal to zero.

$$[x' \; y'] = [x \; y] \begin{bmatrix} a & 0 \\ 0 & 1 \end{bmatrix}$$
$$= [ax \; y]$$

The value of the point is scaled by a factor *a* (Figure 4.3(a)). To scale in the *y* coordinate by *a*, we modify the transformation matrix and obtain:

$$[x' \; y'] = [x \; y] \begin{bmatrix} 1 & 0 \\ 0 & a \end{bmatrix}$$
$$= [x \; ay]$$

This is illustrated in Figure 4.3(b).

We can also scale both x and y simultaneously by the same amount ($a = b$) or by different amounts ($a \neq b$). This is given by:

$$[x' \;\; y'] = [x \;\; y] \begin{bmatrix} a & 0 \\ 0 & b \end{bmatrix}$$

$$= [ax \;\; bx]$$

and is illustrated in Figure 4.3(c).

To obtain a reflection of the point we must change the sign of its corresponding coordinate(s). To reflect about the x axis we change the sign of y, to reflect about the y axis we change the sign of x. Thus:

$$[x' \;\; y'] = [x \;\; y] \begin{bmatrix} 1 & 0 \\ 0 & -1 \end{bmatrix}$$

$$= [x \;\; -y]$$

for x axis reflection and

$$[x' \;\; y'] = [x \;\; y] \begin{bmatrix} -1 & 0 \\ 0 & 1 \end{bmatrix}$$

$$= [-x \;\; y]$$

for y axis reflection. These are illustrated in Figure 4.3(d) and (e) respectively. Once again these two reflections may be combined:

$$[x' \;\; y'] = [x \;\; y] \begin{bmatrix} -1 & 0 \\ 0 & -1 \end{bmatrix}$$

$$= [-x \;\; -y]$$

This will produce a reflection about the origin.

This is illustrated in Figure 4.3(f). You should note from our previous example that only the diagonal terms of the transformation matrix are involved in the operation of scaling, stretching and reflection.

Our final example is the case where an off-diagonal element $a_{21} = 0$, and the main diagonal elements, $a_{11} = a_{22} = 1$. This produces a shear.

$$[x' \;\; y'] = [x \;\; y] \begin{bmatrix} 1 & a \\ 0 & 1 \end{bmatrix}$$

$$= [x \;\; (ax + y)]$$

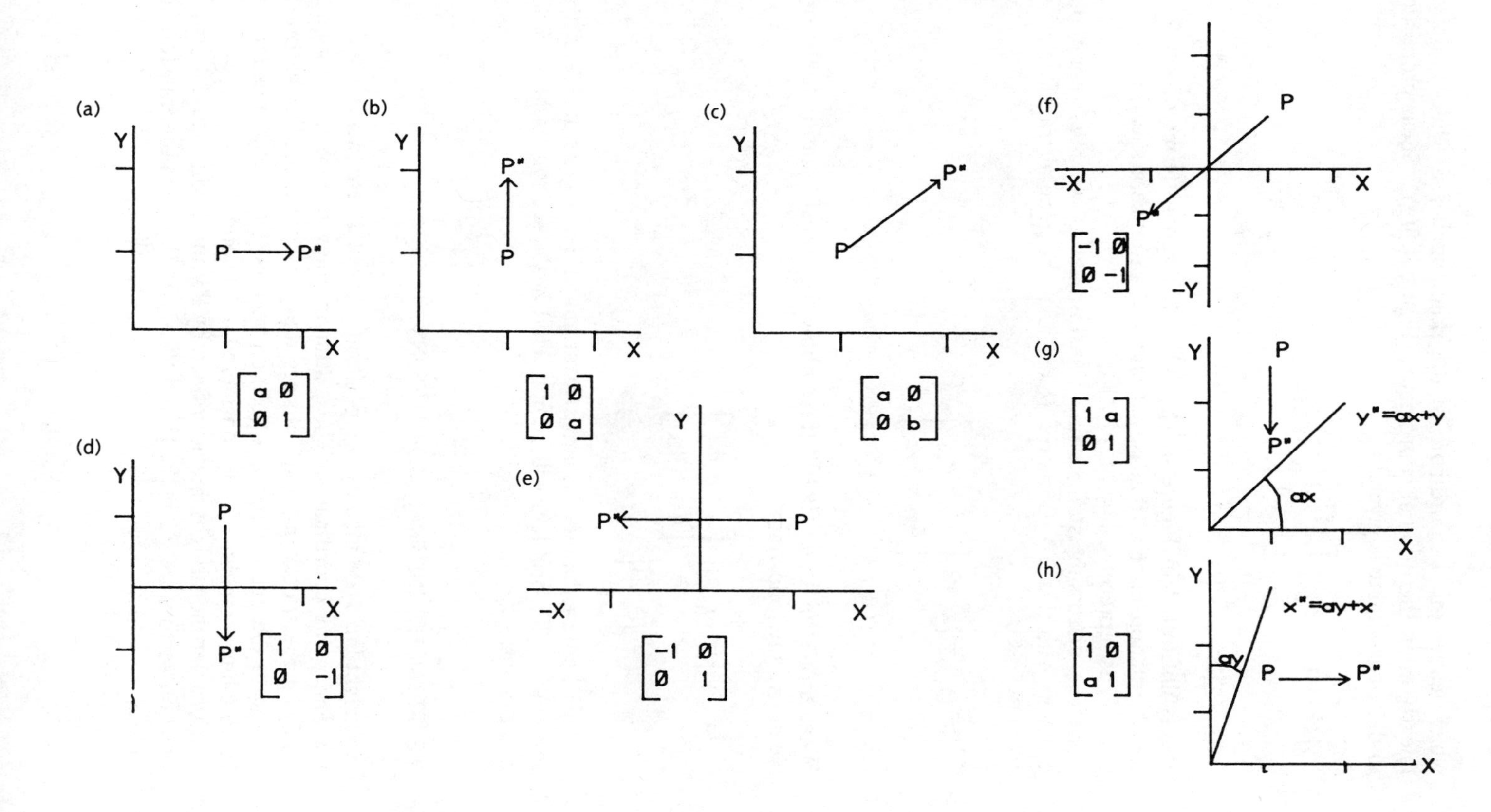

Figure 4.3: Transformations of a Point in Two Dimensions: (a) Scaling in the *x* axis; (b) scaling in the *y* axis; (c) scaling in *x* and *y* axes; (d) *x* axis reflection; (e) *Y* axis reflection; (f) reflection in *x* and *y* axes; (g) shear proportional to the *x* axis; (h) shear proportional to the *Y* axis

which leaves the x coordinate of the point unaltered while y' depends linearly upon the original coordinates (Figure 4.3(g)). Shear proportional to the y coordinate is given by

$$[x' \ \ y'] = [x \ \ y] \begin{bmatrix} 1 & 0 \\ a & 1 \end{bmatrix}$$

$$= [(x + ay) \ \ y]$$

and is illustrated in Figure 4.3(h). The off-diagonal terms of our 2×2 transformation matrix (that is, a_{12} and a_{21}) produce a shearing effect on point P, as represented by $[x \ \ y]$.

Before discussing straight line transformations we must point out a problem with this 2×2 matrix. Recall that the general transformation is given by

$$[x' \ \ y'] = [x \ \ y] \begin{bmatrix} a_{11} & a_{12} \\ a_{21} & a_{22} \end{bmatrix}$$

$$= [(a_{11}x + a_{21}y) \ \ (a_{12}x + a_{22}y)]$$

What about the origin $[0 \ \ 0]$? If we apply this transformation to the origin we derive the following:

$$[x' \ \ y'] = [0 \ \ 0] \begin{bmatrix} a_{11} & a_{12} \\ a_{21} & a_{22} \end{bmatrix}$$

$$= [0 \ \ 0]$$

This implies that the origin is invariant under a general 2×2 transformation. This problem will be corrected in the section on homogeneous coordinates.

4.3 STRAIGHT LINE TRANSFORMATIONS

A straight line is defined by two position vectors. These two vectors specify the end-point coordinates. Thus, straight lines can be transformed by applying the transformation matrices that we considered in the previous section to the line's end-point position vectors. Data structures for representing line segments will be considered in Chapter 7.

Consider the straight line as defined by **P1** and **P2** in Figure 4.4(a). The position vectors are **P1** $= [2 \ \ 1]$ and **P2** $= [3 \ \ 2]$. Given the transformation matrix

Figure 4.4: Straight Line Transformation (Translation) of the Line *P*1 *P*2

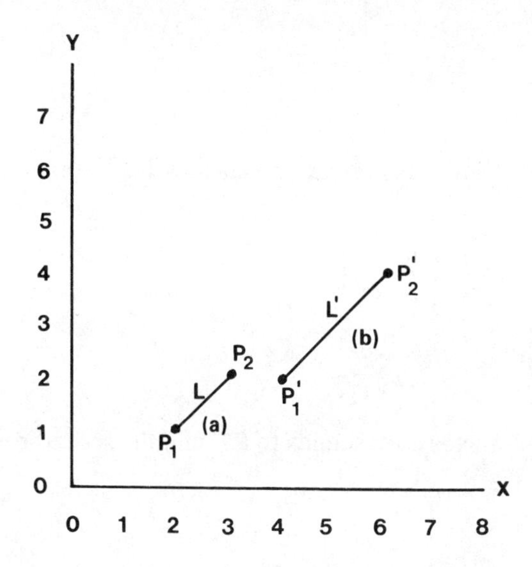

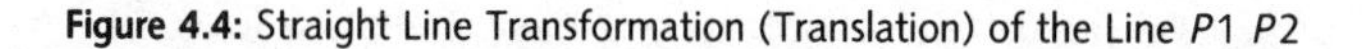

$$\mathbf{T} = \begin{bmatrix} 2 & 0 \\ 0 & 2 \end{bmatrix}$$

what results would we expect? From the previous section we note that both main diagonal elements, that is a_{11}, $a_{22} > 0$, and the off-diagonal elements $a_{12} = a_{21} = 0$. This produces a scale or stretching by a factor of 2 in both the x and y coordinates. We can confirm this by multiplying the position vectors by the transformation matrix **T** and plotting the results. The resultant plot is given in Figure 4.4(b).

$$\mathbf{P1'} = \mathbf{P1T} = [2 \quad 1] \begin{bmatrix} 2 & 0 \\ 0 & 2 \end{bmatrix}$$

$$= [4 \quad 2]$$

and

$$\mathbf{P2'} = \mathbf{P2T} = [3 \quad 2] \begin{bmatrix} 2 & 0 \\ 0 & 2 \end{bmatrix}$$

$$= [6 \quad 4]$$

You will note that we applied the transformation **T** to each end point (position vector) one at a time but in two distinct steps. We can however, take advantage of the matrix structure and construct a matrix of dimension

2 × 2 to contain these two position vectors. In our case it would look like the following:

$$\mathbf{L} = \begin{bmatrix} 2 & 1 \\ 3 & 2 \end{bmatrix}$$

The previous stretching transformation can now be accomplished in one operation:

$$\mathbf{L}' = \mathbf{LT} = \begin{bmatrix} 2 & 1 \\ 3 & 2 \end{bmatrix} \begin{bmatrix} 2 & 0 \\ 0 & 2 \end{bmatrix}$$
$$= \begin{bmatrix} 4 & 2 \\ 6 & 4 \end{bmatrix}$$

where the first row vector corresponds to $\mathbf{P1}'$ and the second row vector to $\mathbf{P2}'$.

4.4 ROTATION

In Figure 4.5(a) we give a plane triangle defined by the points $P1$, $P2$, $P3$. To rotate this object through 90 degrees 'about the origin' in a counter-clockwise manner we can operate on the position vectors with the transformation matrix.

$$\mathbf{T} = \begin{bmatrix} 0 & 1 \\ -1 & 0 \end{bmatrix}$$

As we did with the line, we can use an **N** × 2 matrix to hold the position vector information. With $\mathbf{P1} = [6 \;\; 2]$, $\mathbf{P2} = [4 \;\; 4]$ and $\mathbf{P3} = [8 \;\; 4]$ we get

$$\begin{bmatrix} -2 & 6 \\ -4 & 4 \\ -4 & 8 \end{bmatrix} = \begin{bmatrix} 6 & 2 \\ 4 & 4 \\ 8 & 4 \end{bmatrix} \begin{bmatrix} 0 & 1 \\ -1 & 0 \end{bmatrix}$$

This gives the triangle P' $P2'$ $P3'$ in Figure 4.5(b). For a 180 degree rotation (Figure 4.5(c)

$$\mathbf{T} = \begin{bmatrix} -1 & 0 \\ 0 & -1 \end{bmatrix}$$ which gives

$$\begin{bmatrix} -6 & -2 \\ -4 & -4 \\ -8 & -8 \end{bmatrix} = \begin{bmatrix} 6 & 2 \\ 4 & 4 \\ 8 & 4 \end{bmatrix} \begin{bmatrix} -1 & 0 \\ 0 & -1 \end{bmatrix}$$

and a 270 degree rotation (Figure 4.5(d)) $\mathbf{T} = \begin{bmatrix} 0 & -1 \\ 1 & 0 \end{bmatrix}$ which gives

$$\begin{bmatrix} 2 & -6 \\ 4 & -4 \\ 4 & -8 \end{bmatrix} = \begin{bmatrix} 6 & 2 \\ 4 & 4 \\ 8 & 4 \end{bmatrix} \begin{bmatrix} 0 & -1 \\ 1 & 0 \end{bmatrix}$$

Note that the general 2×2 rotation matrix is given by:

$$\begin{bmatrix} \cos\theta & \sin\theta \\ -\sin\theta & \cos\theta \end{bmatrix}$$

This transformation matrix produces a pure rotation, about the origin, through an arbitrary angle.

Figure 4.5: Counterclockwise Rotation of a Triangle About the Origin

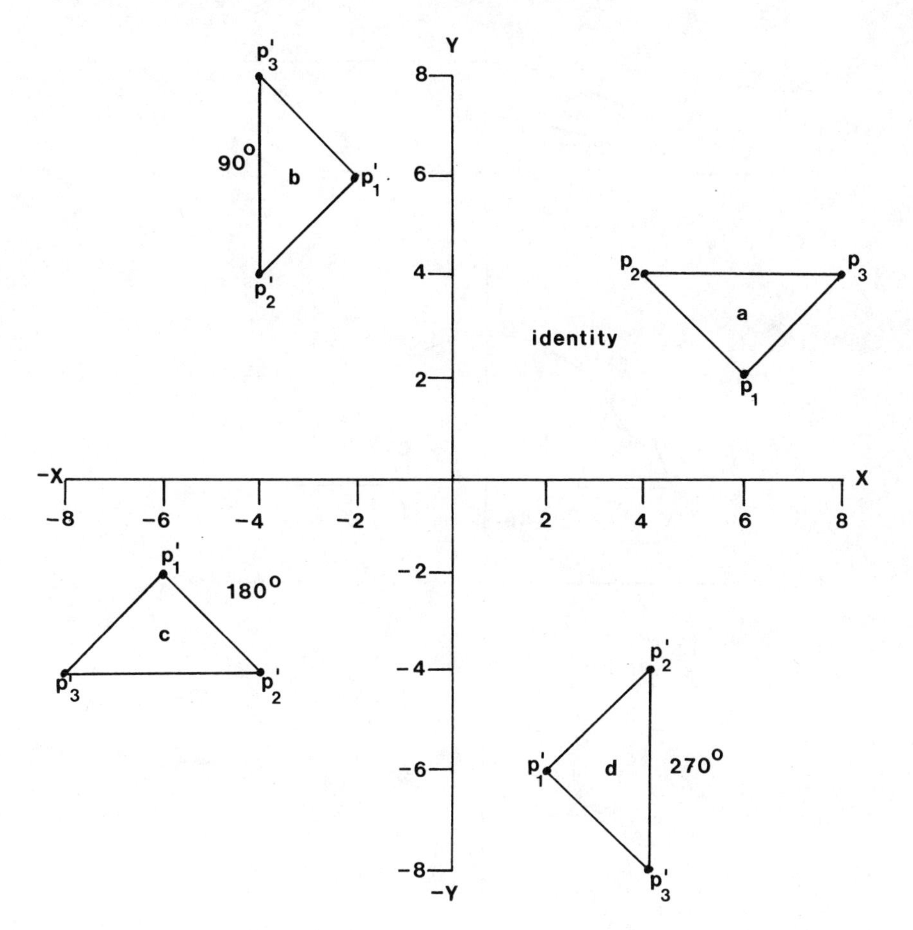

4.5 REFLECTION

Pure two-dimensional rotation in the *xy* plane occurs about an axis normal to the *xy* plane (Figure 4.6(a)) while a reflection is a 180 degree rotation about an axis in the *xy* plane (Figure 4.6(b)).

Figure 4.6: (a) Pure Rotation of a Two-dimensional Object Takes Place in the Two-dimensional Plane Around an Axis Normal to the Plane. (b) Reflections of an Object in Two Dimensions (Here a Triangle) Involve Movement Out of the Plane

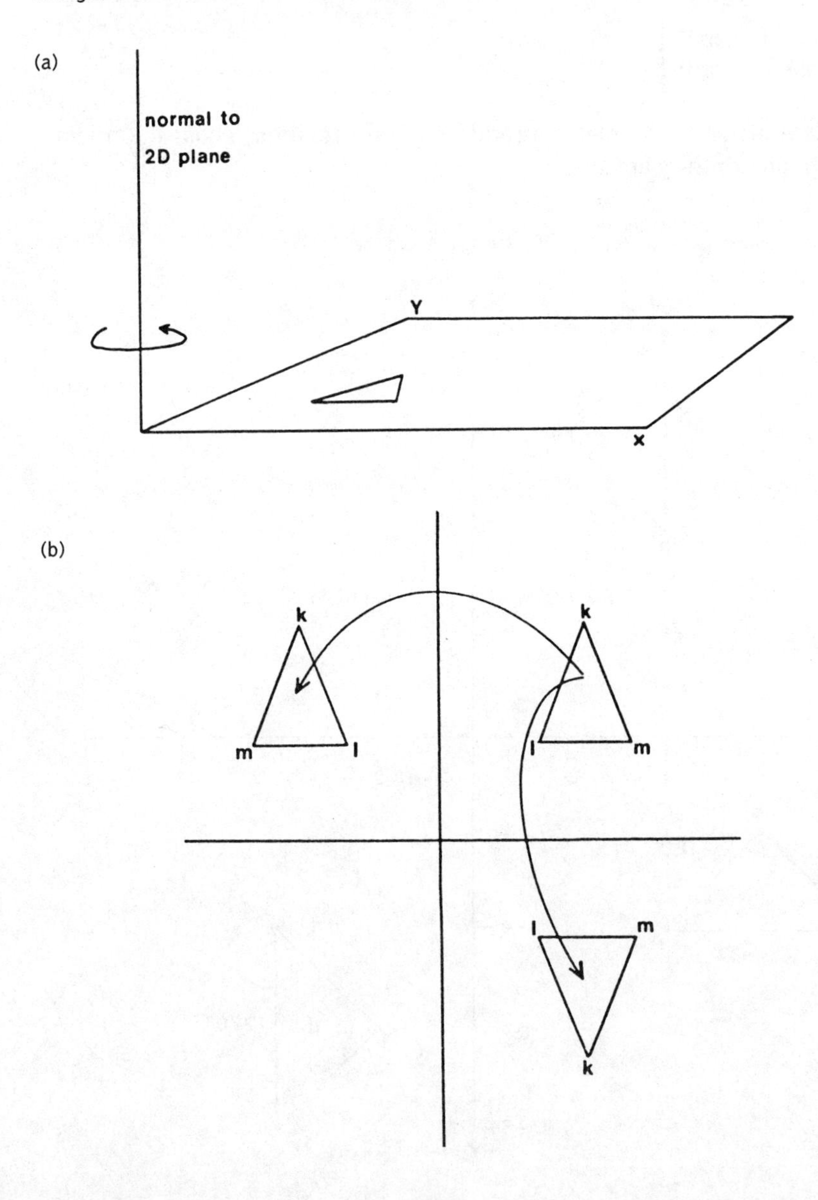

In Figure 4.7 we illustrate a reflection of triangle $P1 = [4 \;\; 2]$, $P2 = [6 \;\; 2]$, $P3 = [6 \;\; 4]$, about the line $y = x$ where the transformation matrix is $\begin{bmatrix} 0 & 1 \\ 1 & 0 \end{bmatrix}$. Again we put the position vectors into a single matrix and get the transformed points:

$$\begin{bmatrix} 2 & 4 \\ 2 & 6 \\ 4 & 6 \end{bmatrix} = \begin{bmatrix} 4 & 2 \\ 6 & 2 \\ 6 & 4 \end{bmatrix} \begin{bmatrix} 0 & 1 \\ 1 & 0 \end{bmatrix}$$

In the above section we have shown how plane surfaces might be manipulated in various ways: reflections, rotations and scaling. We say plane surfaces, as they may be defined by sets of straight lines which themselves are defined by two end vertices. Surface shapes and positions can be controlled through the application of specific matrices to the position vectors which define the vertices. This is fine if only simple single transformations are required, but in general this is not the case. If multiple transformations are desired, there can be a problem since matrix multiplication is, in general 'non-commutative'. That is to say, the order in which the transformations are applied is very important.

Figure 4.7: Reflection of a Triangle About a Line

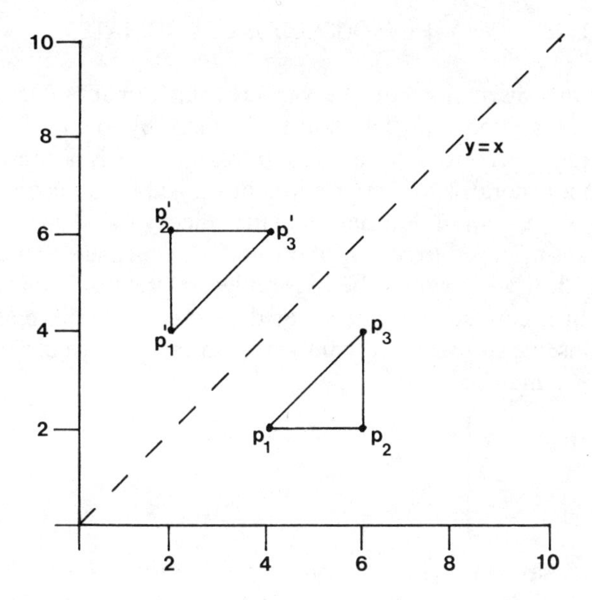

4.6 MULTI-OPERATION TRANSFORMATIONS (COMPOSITION)

We shall now consider an example of the non-commutative nature of matrix multiplication under the operations of rotation and reflection. Consider a reflection about $x=0$ followed by a 90 degree rotation as applied to a point $[x \;\; y]$. The consecutive transformations would produce:

$$\text{reflection} \quad [x' \;\; y'] = [x \;\; y] \begin{bmatrix} -1 & 0 \\ 0 & 1 \end{bmatrix} = [-x \;\; y]$$

$$\text{rotation} \quad [x' \;\; y'] = [-x \;\; y] \begin{bmatrix} 0 & 1 \\ -1 & 0 \end{bmatrix} = [-y \;\; -x]$$

If however we apply the rotation first followed by the reflection we obtain:

$$\text{rotation} \quad [x' \;\; y'] = [x \;\; y] \begin{bmatrix} 0 & 1 \\ -1 & 0 \end{bmatrix} = [-y \;\; x]$$

$$\text{reflection} \quad [x' \;\; y'] = [-y \;\; x] \begin{bmatrix} -1 & 0 \\ 0 & 1 \end{bmatrix} = [y \;\; x]$$

Clearly, $[-y \;\; -x]$ does not equal $[y \;\; x]$. Again great care must be exercised when engaging in matrix composition.

4.7 TWO-DIMENSIONAL HOMOGENEOUS COORDINATES

In the previous discussions on the various transformations that we could apply to points, lines or the objects defined by them, we have not mentioned the operation of translation. In essence this is because within the structure of a general 2×2 matrix it is not possible to accommodate the parameters of translation. Further, in matrix theory translation is treated as an addition, as opposed to scaling and rotations which are treated as multiplications. Clearly it would be desirable to treat all three of these operations in a consistent way so that they may be easily combined. To overcome this, we can add a column vector to the 3×2 matrix to produce a square 3×3 matrix:

$$\begin{bmatrix} 1 & 0 & 0 \\ 0 & 1 & 0 \\ a_{31} & a_{32} & 1 \end{bmatrix}$$

For example, say we wish to translate in the x direction by 5, and the y direction by 3. Then the new position vector would be given by:

$$[x' \ y'] = [x \ y \ 1] \begin{bmatrix} 1 & 0 & 0 \\ 0 & 1 & 0 \\ 5 & 3 & 1 \end{bmatrix} = [x+5 \ \ y+3]$$

The x translation factor is given by a_{31} and the y translation is given by a_{32}. The advantage of expressing position vectors in homogeneous coordinates is that now all three transformations can be treated as multiplications.

Let us consider an example transformation on straight line data. The digitized outlines of a number of sectioned cells have been stored in a computer file as a list of x,y coordinates and associated line segments. The data are centred around the origin for each cell, and a non-overlapping display of the cells is required for comparative purposes (Figure 4.8). Translation of the data is therefore required so that the cells can be displayed side by side. If we assume that the programmer selects appropriate x,y coordinates for centring each cell outline, the translation for a single cell to be centred around the location $x = 50$ and $y = 30$ may be performed in the following manner. Because the original data were centred around the origin, each data point is translated by simply specifying the appropriate translation values in the 3×3 transformation matrix. In the present example the transformation would be as follows:

$$[x' \ y' \ 1] = [x \ \ y \ \ 1] \begin{bmatrix} 1 & 0 & 0 \\ 0 & 1 & 0 \\ 50 & 30 & 1 \end{bmatrix}$$

In homogeneous coordinates, the point given by $[x \ y]$ is represented as $[\mathrm{H}x \ \mathrm{H}y \ \mathrm{H}]$ for any scale factor $\mathrm{H} \neq 0$. If $x^* = \mathrm{H}x$ and $y^* = \mathrm{H}y$, then given any point $\mathbf{P}(x^* \ \ y^* \ \ \mathrm{H})$, we can determine the cartesian coordinate representation for the point as $x = x^*/\mathrm{H}$ and $y = y^*/\mathrm{H}$. In the present context of two-dimensional transformations H will always be 1.

Let us now consider an example of transformation composition which illustrates the beauty of homogeneous coordinates. First recall that all of our previous transformations were defined relative to the origin. What if, as is usually the case, we wish to rotate an object about some arbitrary point p_1? To effect this we must first translate p_1 to the origin, effect the transformation(s), and then translate the point at the origin back to the point p_1. An illustration of this is given in Figure 4.9. If $p_1=(x_1 \ y_1)$ then the complete sequence of transformations looks like this:

$$\begin{bmatrix} 1 & 0 & 0 \\ 0 & 1 & 0 \\ -x_1 & -y_1 & 1 \end{bmatrix} \begin{bmatrix} \cos\theta & \sin\theta & 0 \\ -\sin\theta & \cos\theta & 0 \\ 0 & 0 & 1 \end{bmatrix} \begin{bmatrix} 1 & 0 & 0 \\ 0 & 1 & 0 \\ x_1 & y_1 & 1 \end{bmatrix}$$

Figure 4.8: Translation of Biological Data. Outlines of six cells have been digitized around the origin (a). Translation allows the cell outlines to be compared on the same drawing (b)

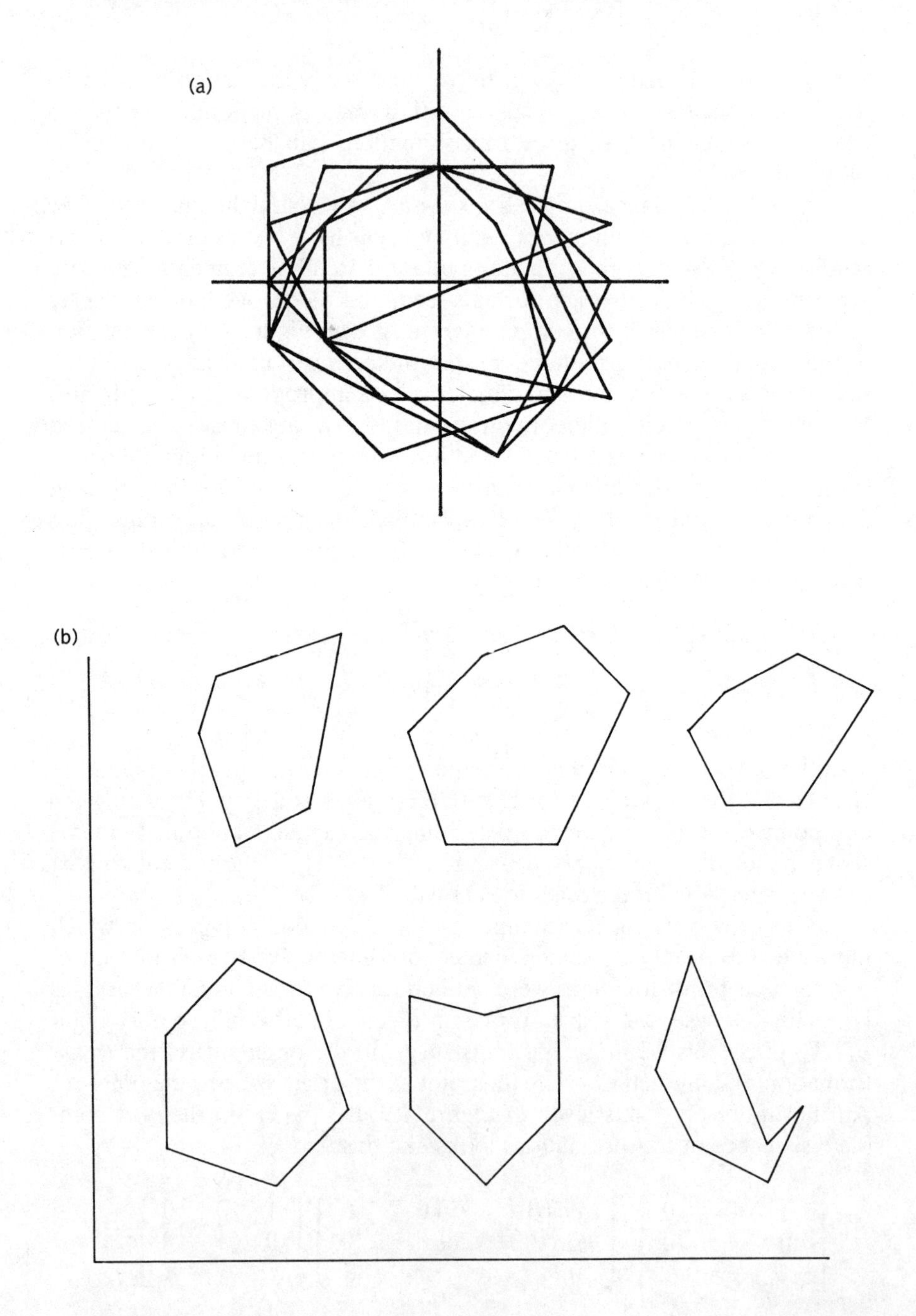

Figure 4.9: An Example of Composite Transformation: Rotation around an Arbitrary Point $p1$. (a) Start position for triangle; (b) translation to the origin; (c) rotation about the origin; (d) translation back to original position

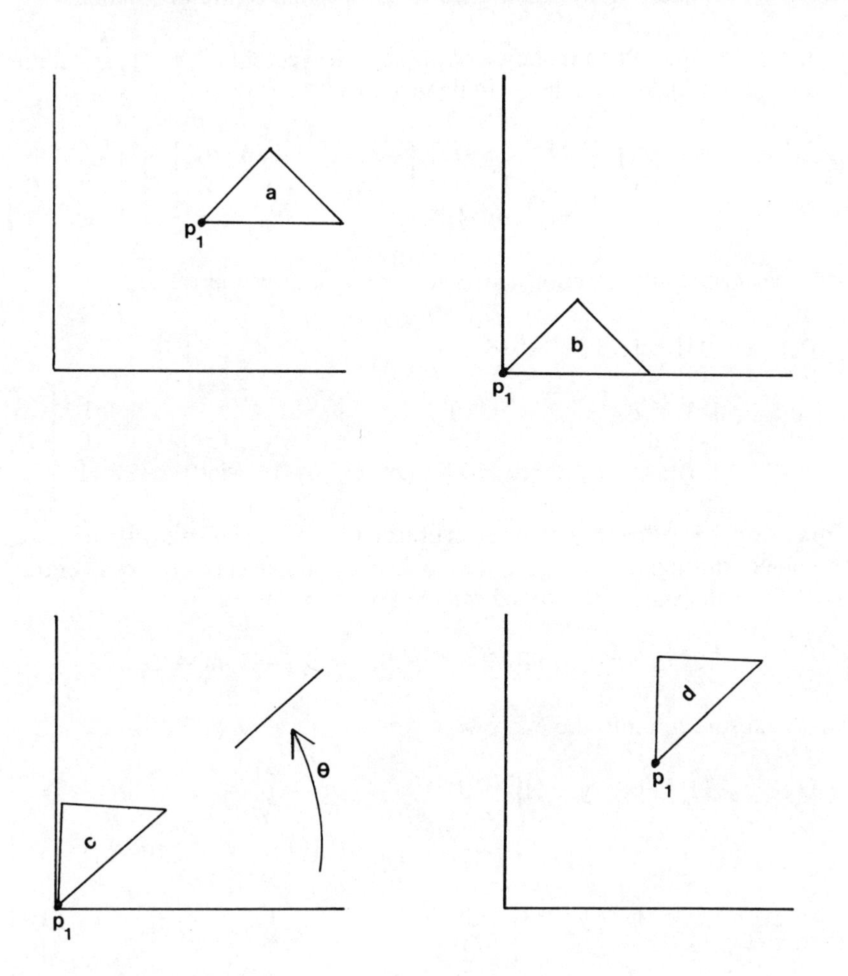

4.8 TWO-DIMENSIONAL ROTATION ABOUT AN ARBITRARY AXIS

In this chapter, all the rotations we have considered have occurred about the origin. However, homogeneous coordinates provide a mechanism for performing rotations about an axis (a single point in the plane) other than the origin. For the most general case, a rotation about an arbitrary point can be produced by:

(1) Translating the centre of rotation (a point) to the origin.
(2) Applying the desired rotation.
(3) Translation of the result back to the original centre of rotation.

In matrix notation, the rotation of the position vector [x y 1] about the point 1,*m* through an angle θ will be given by:

$$[x'\ y'\ H] = [x\ y\ 1] \begin{bmatrix} 1 & 0 & 0 \\ 0 & 1 & 0 \\ -l & -m & 1 \end{bmatrix} \begin{bmatrix} \cos\theta & \sin\theta & 0 \\ -\sin\theta & \cos\theta & 0 \\ 0 & 0 & 1 \end{bmatrix} \begin{bmatrix} 1 & 0 & 0 \\ 1 & 0 & 0 \\ l & m & 1 \end{bmatrix}$$

after performing the interior matrix multiplications we derive:

$$[x'\ \ y'\ \ H] = [x\ \ y\ \ 1] \times \begin{bmatrix} \mathrm{Cos}\,\theta & \mathrm{Sin}\,\theta & 0 \\ -\mathrm{Sin}\,\theta & \mathrm{Cos}\,\theta & 0 \\ (m\sin\theta + l(1-\cos\theta)) & (m(1-\cos\theta) - l\sin\theta) & 1 \end{bmatrix}$$

For example, say that the centre of an object is at [5 2] and we wish to rotate the object through 90° in a counterclockwise manner about its own central axis. The values that are needed for the above expression are:

$$\mathrm{COS}\,\theta = \mathrm{COS}\,90° = 0,\ \mathrm{SIN}\,\theta = \mathrm{SIN}\,90° = 1,\ l = 5,\ \mathrm{m} = 2.$$

Upon substitution into the expression we get:

$$[x'\ \ y'\ \ H] = [x\ \ y\ \ 1] \begin{bmatrix} 0 & 1 & 0 \\ -1 & 0 & 0 \\ (-5(0-1)+2(1)) & (-5(1)-2(0-1)) & 1 \end{bmatrix}$$

$$= [x\ \ y\ \ 1] \begin{bmatrix} 0 & 1 & 0 \\ -1 & 0 & 0 \\ 7 & -3 & 1 \end{bmatrix}$$

then, $x^* = x'/H$, $y^* = y'/H$

In Table 4.1 we present a summary of the general 2 × 2 transformations, excluding rotations.

Table 4.1: General 2 × 2 Transformations Excluding the Rotations

Given a position vector $\mathbf{P} = [x \quad y]$ and a general transformation matrix

$\mathbf{T} = \begin{bmatrix} a_{11} & a_{12} \\ a_{21} & a_{22} \end{bmatrix}$ we can define the following:

Identity: $\mathbf{T} = \begin{bmatrix} 1 & 0 \\ 0 & 1 \end{bmatrix}$ $[x' \quad y'] = \mathbf{PT} = [x \quad y] \begin{bmatrix} 1 & 0 \\ 0 & 1 \end{bmatrix} = [x \quad y]$

Identity transformation. No change in point position.

$x \rightarrow x;\ y \rightarrow y.$

Reflection: $\mathbf{T} = \begin{bmatrix} -1 & 0 \\ 0 & 1 \end{bmatrix}$ $[x' \quad y'] = \mathbf{PT} = [x \quad y] \begin{bmatrix} -1 & 0 \\ 0 & 1 \end{bmatrix} = [-x \quad y]$

Reflection of point about y axis. $x \rightarrow -x;\ y \rightarrow y$

$\mathbf{T} = \begin{bmatrix} 1 & 0 \\ 0 & -1 \end{bmatrix}$ $[x' \quad y'] = \mathbf{PT} = [x \quad y] \begin{bmatrix} 1 & 0 \\ 0 & -1 \end{bmatrix} = [x \quad y]$

Reflection of point about x axis. $x \rightarrow x;\ y \rightarrow -y$

$\mathbf{T} = \begin{bmatrix} -1 & 0 \\ 0 & -1 \end{bmatrix}$ $[x' \quad y'] = PT = [x \quad y] \begin{bmatrix} -1 & 0 \\ 0 & -1 \end{bmatrix} = [-x \quad -y]$

Reflection of point through origin. $x \rightarrow -x;\ y \rightarrow -y$

Let $a > 1$, $b > 1$

Scale: $\mathbf{T} = \begin{bmatrix} a & 0 \\ 0 & 1 \end{bmatrix}$ $[x' \quad y'] = \mathbf{PT} = [x \quad y] \begin{bmatrix} a & 0 \\ 0 & 1 \end{bmatrix} = [ax \quad y]$

Scale or stretching of x axis by factor of a.

$x \rightarrow ax;\ y \rightarrow y$

$\mathbf{T} = \begin{bmatrix} 1 & 0 \\ 0 & a \end{bmatrix}$ $[x' \quad y'] = \mathbf{PT} = [x \quad y] \begin{bmatrix} 1 & 0 \\ 0 & a \end{bmatrix} = [x \quad ay]$

Scale or stretching of y axis by factor of a.

$x \rightarrow x;\ y \rightarrow ay$

$\mathbf{T} = \begin{bmatrix} a & 0 \\ 0 & b \end{bmatrix}$ $[x' \quad y'] = \mathbf{PT} = [x \quad y] \begin{bmatrix} a & 0 \\ 0 & b \end{bmatrix} = [ax \quad by]$

Scale or stretching of x by a and a scale or stretching of y by a factor of b. If $a = b$ then no distortion of figure, if $a \neq b$ then distortion of figure. $x \rightarrow ax;\ y \rightarrow by$

Shear: $\mathbf{T} = \begin{bmatrix} 1 & 0 \\ a & 1 \end{bmatrix}$ $[x' \quad y'] = \mathbf{PT} = [x \quad y] \begin{bmatrix} 1 & 0 \\ a & 1 \end{bmatrix} = [x + ay \quad y]$

Shear proportional to the y coordinate.

$x \rightarrow x + ay;\ y \rightarrow y$

Table 4.1 continued

$$\mathbf{T} = \begin{bmatrix} 1 & a \\ 0 & 1 \end{bmatrix} \quad [x' \ \ y'] = \mathbf{PT} = [x \ \ y] \begin{bmatrix} 1 & a \\ 0 & 1 \end{bmatrix} = [x \ \ ax + y]$$

Shear proportional to the x coordinate.

$x \rightarrow x$; $y \rightarrow ax + y$

$$\mathbf{T} = \begin{bmatrix} 1 & a \\ b & 1 \end{bmatrix} \quad [x' \ \ y'] = \mathbf{PT} = [x \ \ y] \begin{bmatrix} 1 & a \\ b & 1 \end{bmatrix} = [x + by \ \ y + ax]$$

Shear proportional to both the x and y coordinates.

$x \rightarrow x + by$; $y \rightarrow y + ax$

4.9 REFERENCES

Angell, I.O. *A Practical Introduction to Computer Graphics* (Macmillan Press, London, 1981)
Bowyer, A. and Woodwark, J. *A Programmer's Geometry* (Butterworths, London, 1983)
Newman, W.M. and Sproull, R.F. *Principles of Interactive Computer Graphics*, 2nd ed (McGraw-Hill, New York, 1979)
Rogers, D.F. and Adams, J.A. *Mathematical Elements for Computer Graphics* (McGraw-Hill, Maidenhead, Berkshire, 1976)

5 Three-dimensional Graphics

5.1 BASIC CONCEPTS

In Chapter 4 we considered the basic transformations that are available to us for manipulating planar figures. We now enter the realm of three-dimensional objects in three-dimensional space. This chapter is primarily concerned with the transformations which are applicable to these objects. While most of the information that we discussed in Chapter 4 is relevant to the three-dimensional case, there are substantial differences. These differences arise mainly through the addition of a third axis (at right angles to the existing two). The transformations are similar, but the complexity has increased.

We begin our discussion of these three-dimensional transformations by first considering a few basic concepts. Our first decision must be the choice of the rectangular coordinate system. The reader should, however, note that there are several other coordinate systems available for the purposes of modelling, for example, cylindrical or spherical. However, from a graphics point of view there is only the rectangular system. Therefore, if you are working in systems of coordinates other than rectangular, you must eventually transform them back to rectangular coordinates for computer graphical display. Recall that in the two-dimensional case we had only the simple x,y cartesian coordinate system available, as illustrated in Figure 5.1. In the three-dimensional case we have two choices with respect to the placement of this third axis, in other words, the z coordinate axis. These two systems are known as the right-hand system (RHS) and the left-hand system (LHS). They are illustrated in Figure 5.2(a) and Figure 5.2(b) respectively. The first point to observe is that, as our right and left hands are mirror images of each other, so too are these coordinate systems.

The choice of which system to use in your computer graphics applications is largely arbitrary as long as one is consistent throughout a given application. The advantages or disadvantages of one system over the other are, in general, a matter of personal preference. The RHS is the more common in everyday use, while the LHS 'seems' to be more suitable to the graphics environment. This suitability stems from the fact that if you imagine the LHS system mapped onto the surface of the display screen, with the x axis running horizontally along the bottom and the y axis vertical along the left side, the z axis values increase as one moves away from the observer.

Once we have chosen a coordinate system, several parameters become fixed. One of these is the direction of positive or negative rotation about a given axis. A simple method for determining this positive rotation is as follows: consider the LHS first. Hold your left hand in front of you with

Figure 5.1: Simple *X,Y* Cartesian Coordinate System in Two Dimensions

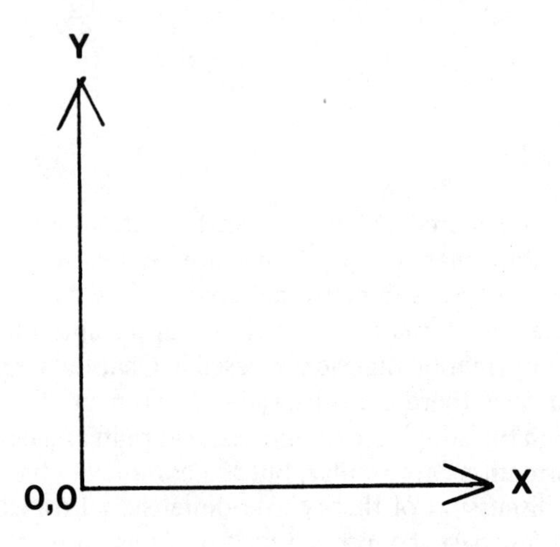

Figure 5.2: Three-dimensional Coordinate Systems: (a) Right-hand System; (b) Left-hand System

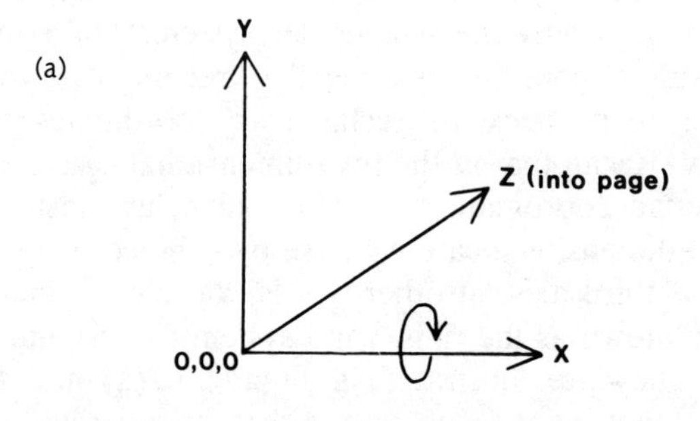

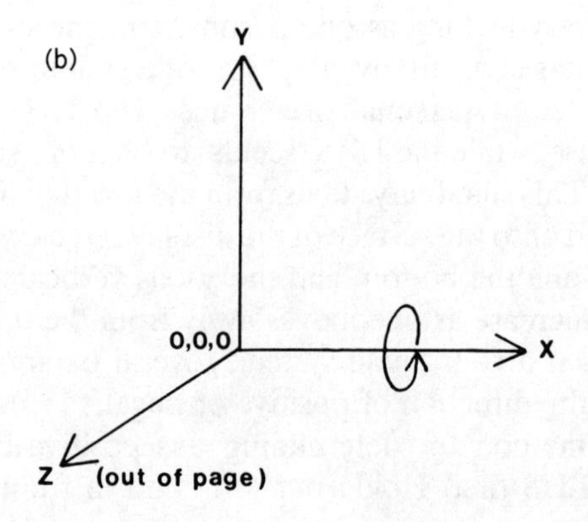

your thumb pointing at your nose. If you imagine a positive value of the axis at your eye and the origin of the coordinate system pointing away from you, the 'curl' of your fingers gives the direction of positive rotation. The same is true for the RHS, but of course, you use your right hand. You should observe that a clockwise rotation about an axis in one system is a counterclockwise rotation in the other.

5.2 THREE-DIMENSIONAL HOMOGENEOUS COORDINATES

One of the most important concepts introduced in Chapter 4 was the notion of homogeneous coordinates. Such coordinates are represented in three-dimensional space by a four-dimensional position vector $[x \quad y \quad z \quad 1]$ or $[\mathrm{X} \quad \mathrm{Y} \quad \mathrm{Z} \quad \mathrm{H}]$. Recall that the transformation from homogeneous coordinates to ordinary coordinates is given by:

$$[\mathrm{X} \quad \mathrm{Y} \quad \mathrm{Z} \quad \mathrm{H}] = [x \quad y \quad z \quad 1]\mathbf{A}$$

and

$$[x' \quad y \quad z' \quad 1] = \left[\frac{\mathrm{X}}{\mathrm{H}} \quad \frac{\mathrm{Y}}{\mathrm{H}} \quad \frac{\mathrm{Z}}{\mathrm{H}} \quad 1\right]$$

where **A** is some transformation matrix. The general 4 × 4 transformation matrix for three-dimensional homogeneous coordinates is:

$$\mathbf{A} = \begin{bmatrix} a_{11} & a_{12} & a_{13} & a_{14} \\ a_{21} & a_{22} & a_{23} & a_{24} \\ a_{31} & a_{32} & a_{33} & a_{34} \\ a_{41} & a_{42} & a_{43} & a_{44} \end{bmatrix}$$

We will find it convenient to partition this 4 × 4 matrix into four distinct submatrices:

$$\left[\begin{array}{c|c} 3 \times 3 & \begin{matrix} 3 \\ \times \\ 1 \end{matrix} \\ \hline 1 \times 3 & 1 \times 1 \end{array}\right]$$

These four submatrices produce the following effects: The 3 × 3 matrix affects the scaling, shearing and rotation. The 1 × 3 row matrix affects translation. The 3 × 1 column matrix affects the perspective and the single element (1 × 1) affects the overall scaling. Thus, with this one matrix and suitable entries, we can produce shearing, local scaling, rotation, reflection, translation, perspective and overall scaling.

5.3 THREE-DIMENSIONAL SCALING

The main diagonal terms of this general 4 × 4 transformation matrix produce local and overall scaling. Consider the following:

$$[x' \quad y' \quad z' \quad 1] = [x \quad y \quad z \quad 1] \begin{bmatrix} a_{11} & 0 & 0 & 0 \\ 0 & a_{22} & 0 & 0 \\ 0 & 0 & a_{33} & 0 \\ 0 & 0 & 0 & 1 \end{bmatrix}$$

$$= [a_{11}x \quad a_{22}y \quad a_{33}z \quad 1]$$

which illustrates local scaling. The scale of the x component is determined by a_{11}, the y component by a_{22} and the z component by a_{33}. To affect the overall scaling we utilise the fourth main diagonal element, a_{44}. Consider the following:

$$[X \quad Y \quad Z \quad H] = [x \quad y \quad z \quad 1] \begin{bmatrix} 1 & 0 & 0 & 0 \\ 0 & 1 & 0 & 0 \\ 0 & 0 & 1 & 0 \\ 0 & 0 & 0 & a_{44} \end{bmatrix}$$

$$= [x \quad y \quad z \quad a_{44}]$$

Therefore from §5.2

$$[x' \quad y' \quad z' \quad 1] = \begin{bmatrix} \frac{x}{a_{44}} & \frac{y}{a_{44}} & \frac{z}{a_{44}} & 1 \end{bmatrix}$$

We note that the same effect of overall scaling may be obtained by means of equal local scalings. If $a_{44} = s$ then by taking instead $a_{11} = a_{22} = a_{33} = 1/s$ and $a_{44} = 1$ we obtain the same scaling effect.

5.4 THREE-DIMENSIONAL SHEARING

Shear in three dimensions is produced by utilising the off-diagonal elements in the upper left 3 × 3 submatrix of the 4 × 4 general transformation matrix. For example:

$$[x' \quad y' \quad z' \quad 1] = [x \quad y \quad z \quad 1] \begin{bmatrix} 1 & a_{12} & a_{13} & 0 \\ a_{21} & 1 & a_{23} & 0 \\ a_{31} & a_{32} & 1 & 0 \\ 0 & 0 & 0 & 1 \end{bmatrix}$$

$$= [x + a_{21}y + a_{31}z \quad a_{12}x + y + a_{32}z \quad a_{13}x + a_{23}y + z \quad 1]$$

5.5 THREE-DIMENSIONAL ROTATIONS

We have previously seen that the 3 × 3 submatrix could affect the operation of shear and scaling. If, however the determinant of this submatrix is +1, then the effect is a pure rotation about this origin. There are three special cases that we shall consider prior to describing three-dimensional rotation about an arbitrary axis.

The first case that we consider is a rotation about the x axis. In this transformation, the x components do not vary. This would imply that the first row and first column of the transformation matrix would have elements of zero, except for unity on the main diagonal (a_{11}). Rotation will be assumed to be positive as in a right-handed system. For rotation through an angle θ (theta) about the x axis the transformation matrix **A** is given by:

$$\mathbf{A} = \begin{bmatrix} 1 & 0 & 0 & 0 \\ 0 & \cos\theta & \sin\theta & 0 \\ 0 & -\sin\theta & \cos\theta & 0 \\ 0 & 0 & 0 & 1 \end{bmatrix}$$

In the case of a rotation of angle φ (phi) about the y axis, the second row and second column of the transformation matrix contain elements of zero, except for unity on the main diagonal ($a_{22} = 1$). The transformation matrix is given by:

$$\mathbf{A} = \begin{bmatrix} \cos\varphi & 0 & -\sin\varphi & 0 \\ 0 & 1 & 0 & 0 \\ \sin\varphi & 0 & \cos\varphi & 0 \\ 0 & 0 & 0 & 1 \end{bmatrix}$$

A rotation of angle ψ (psi) about the z axis is defined in a similar

$$\mathbf{A} = \begin{bmatrix} \cos\psi & \sin\psi & 0 & 0 \\ -\sin\psi & \cos\psi & 0 & 0 \\ 0 & 0 & 1 & 0 \\ 0 & 0 & 0 & 1 \end{bmatrix}$$

Three-dimensional rotations are produced by matrix multiplication and as such are non-commutative. That is to say, the order of multiplication is important. To illustrate this consider a rotation about the x axis followed by an equal rotation about the z axis.

$$\mathbf{A} = \mathbf{XZ}$$

$$= \begin{bmatrix} 1 & 0 & 0 & 0 \\ 0 & \cos\theta & \sin\theta & 0 \\ 0 & -\sin\theta & \cos\theta & 0 \\ 0 & 0 & 0 & 1 \end{bmatrix} \begin{bmatrix} \cos\theta & \sin\theta & 0 & 0 \\ -\sin\theta & \cos\theta & 0 & 0 \\ 0 & 0 & 1 & 0 \\ 0 & 0 & 0 & 1 \end{bmatrix}$$

$$= \begin{bmatrix} \cos\theta & \sin\theta & 0 & 0 \\ -\sin\theta\cos\theta & \cos^2\theta & \sin\theta & 0 \\ \sin^2\theta & -\sin\theta\cos\theta & \cos\theta & 0 \\ 0 & 0 & 0 & 1 \end{bmatrix}$$

The reverse operation of a rotation about the z axis followed by an equal rotation about the x axis yields:

$$\mathbf{B} = \mathbf{ZX}$$

$$= \begin{bmatrix} \cos\theta & \sin\theta & 0 & 0 \\ -\sin\theta & \cos\theta & 0 & 0 \\ 0 & 0 & 1 & 0 \\ 0 & 0 & 0 & 1 \end{bmatrix} \begin{bmatrix} 1 & 0 & 0 & 0 \\ 0 & \cos\theta & \sin\theta & 0 \\ 0 & -\sin\theta & \cos\theta & 0 \\ 0 & 0 & 0 & 1 \end{bmatrix}$$

$$= \begin{bmatrix} \cos\theta & \sin\theta\cos\theta & \sin^2\theta & 0 \\ -\sin\theta & \cos^2\theta & \sin\theta\cos\theta & 0 \\ 0 & -\sin\theta & \cos\theta & 0 \\ 0 & 0 & 0 & 1 \end{bmatrix}$$

Clearly, $\mathbf{A} \neq \mathbf{B}$

5.6 REFLECTION IN THREE DIMENSIONS

There are occasions when a reflection of a three-dimensional object is required. For example, when 180 degree rotations are needed about a principal axis. In this case a reflection would require less computation than a full rotation using the trigonometric functions in the 4×4 transformation matrix. As a final example we could use a reflection in changing from a RHS to a LHS or vice versa.

The simplest reflections, in three dimensions, occur through a plane. For a rigid-body reflection the determinant of the transformation matrix will be -1.0. The most straightforward way to achieve this is to change the sign of the coordinate not affected by the transformation. Table 5.1 below illustrates this for reflection through the yz, xz, xy planes. This corresponds to changing the sign for the x, y, z coordinates respectively.

Table 5.2: Three-dimensional Reflection through Planes *yz*, *xz*, *xy* Achieved by Changing the Sign of the *x*, *y* and *z* Coordinates Respectively

Reflection through plane	Change sign of coordinate	Transformation matrix to perform reflection
yz	*x*	$\begin{bmatrix} -1 & 0 & 0 & 0 \\ 0 & 1 & 0 & 0 \\ 0 & 0 & 1 & 0 \\ 0 & 0 & 0 & 1 \end{bmatrix}$
xz	*y*	$\begin{bmatrix} 1 & 0 & 0 & 0 \\ 0 & -1 & 0 & 0 \\ 0 & 0 & 1 & 0 \\ 0 & 0 & 0 & 1 \end{bmatrix}$
xy	*z*	$\begin{bmatrix} 1 & 0 & 0 & 0 \\ 0 & 1 & 0 & 0 \\ 0 & 0 & -1 & 0 \\ 0 & 0 & 0 & 1 \end{bmatrix}$

5.7 THREE-DIMENSIONAL TRANSLATION

As stated previously, the 1×3 row vector, $[a_{41} \; a_{42} \; a_{43}]$ controls the *x*, *y*, *z* translation factors. The transformation is given by:

$$[\mathrm{X} \quad \mathrm{Y} \quad \mathrm{Z} \quad \mathrm{H}] = [x \quad y \quad z \quad 1] \begin{bmatrix} 1 & 0 & 0 & 0 \\ 0 & 1 & 0 & 0 \\ 0 & 0 & 1 & 0 \\ a_{41} & a_{42} & a_{43} & 1 \end{bmatrix}$$

upon expansion we get:

$$[\mathrm{X} \quad \mathrm{Y} \quad \mathrm{Z} \quad \mathrm{H}] = [(x + a_{41}) \quad (y + a_{42}) \quad (z + a_{43}) \quad 1]$$

5.8 THREE-DIMENSIONAL ROTATIONS ABOUT AN ARBITRARY AXIS

In Chapter 4 we discussed the procedure for two-dimensional plane rotation about an arbitrary axis. We will now generalise the problem for rotation about any axis in three-dimensional space. As in the two-dimensional case the procedure requires:

(1) A three-dimensional translation (the object and the desired axis of

rotation such that the rotation is made about an axis passing through the origin of the coordinate system).

(2) A rotation about the origin.

(3) And a three-dimensional translation back from the origin.

If the axis of the required rotation passes through the point $P = [a \quad b \quad c \quad 1]$, the matrix description of the transformation is given by:

$$[X \quad Y \quad Z \quad H]$$

$$= [x \quad y \quad z \quad 1] \begin{bmatrix} 1 & 0 & 0 & 0 \\ 0 & 1 & 0 & 0 \\ 0 & 0 & 1 & 0 \\ -a & -b & -c & 1 \end{bmatrix} [\mathbf{R}] \begin{bmatrix} 1 & 0 & 0 & 0 \\ 0 & 1 & 0 & 0 \\ 0 & 0 & 1 & 0 \\ a & b & c & 1 \end{bmatrix}$$

The elements of the 4 × 4 general rotation matrix **R** are given by:

$$\begin{pmatrix} n_1^2+(1-n_1^2)\cos\theta & n_1n_2(1-\cos\theta)+n_3\sin\theta & n_1n_3(1-\cos\theta)-n_2\sin\theta & 0 \\ n_1n_2(1-\cos\theta)-n_3\sin\theta & n_2^2+(1-n_2^2)\cos\theta & n_2n_3(1-\cos\theta)+n_1\sin\theta & 0 \\ n_1n_3(1-\cos\theta)+n_2\sin\theta & n_2n_3(1-\cos\theta)-n_1\sin\theta & n_3^2+(1-n_3^2)\cos\theta & 0 \\ 0 & 0 & 0 & 1 \end{pmatrix}$$

Where n_1, n_2 and n_3 are the direction cosines.

The three matrices given earlier in this chapter for rotation about the x, y, z axes are special cases of this general form. Consider an arbitrary axis of rotation (ON) translated such that it passes through the origin, as illustrated in Figure 5.3. We next rotate a point on the translated object, e.g. P, about the axis by an angle θ. The effect of this is to move the point from P to P'. In Figure 5.3, $PQ = P'Q$, with both PQ and $P'Q$ perpendicular to OQ, $P'S$ is constructed to be perpendicular to PQ. The determination of the elements in the rotation matrix, **R**, is performed by relating the transformed coordinates of P' to:

(1) The coordinates of P

(2) The rotation angle θ

(3) The direction of the axis of rotation as specified by the unit vector **n**.

The direction of the axis of rotation can be expressed in terms of its three direction cosines, $n_1 = \cos\alpha$, $n_2 = \cos\beta$, $n_3 = \cos\gamma$ which form the unit vector **n**. The angles α, β and γ are illustrated in Figure 5.3(b). For a general vector $\mathbf{Q} = q_1\mathbf{i} + q_2\mathbf{i} + q_3\mathbf{k}$, the unit vector is calculated by $\mathbf{n} = \mathbf{Q} / |\mathbf{Q}|$, where $|\mathbf{Q}|$ is the absolute value of the vector and is given by

$$|\mathbf{Q}| = (q_1^2 + q_2^2 + q_3^2)^{1/2}$$

Figure 5.3: (a) Three-dimensional Rotation about an Arbitrary Axis. See Text for Description. (b) Illustration of Angles *a*, *b* and *c* Used in the Calculation of the Direction Cosines

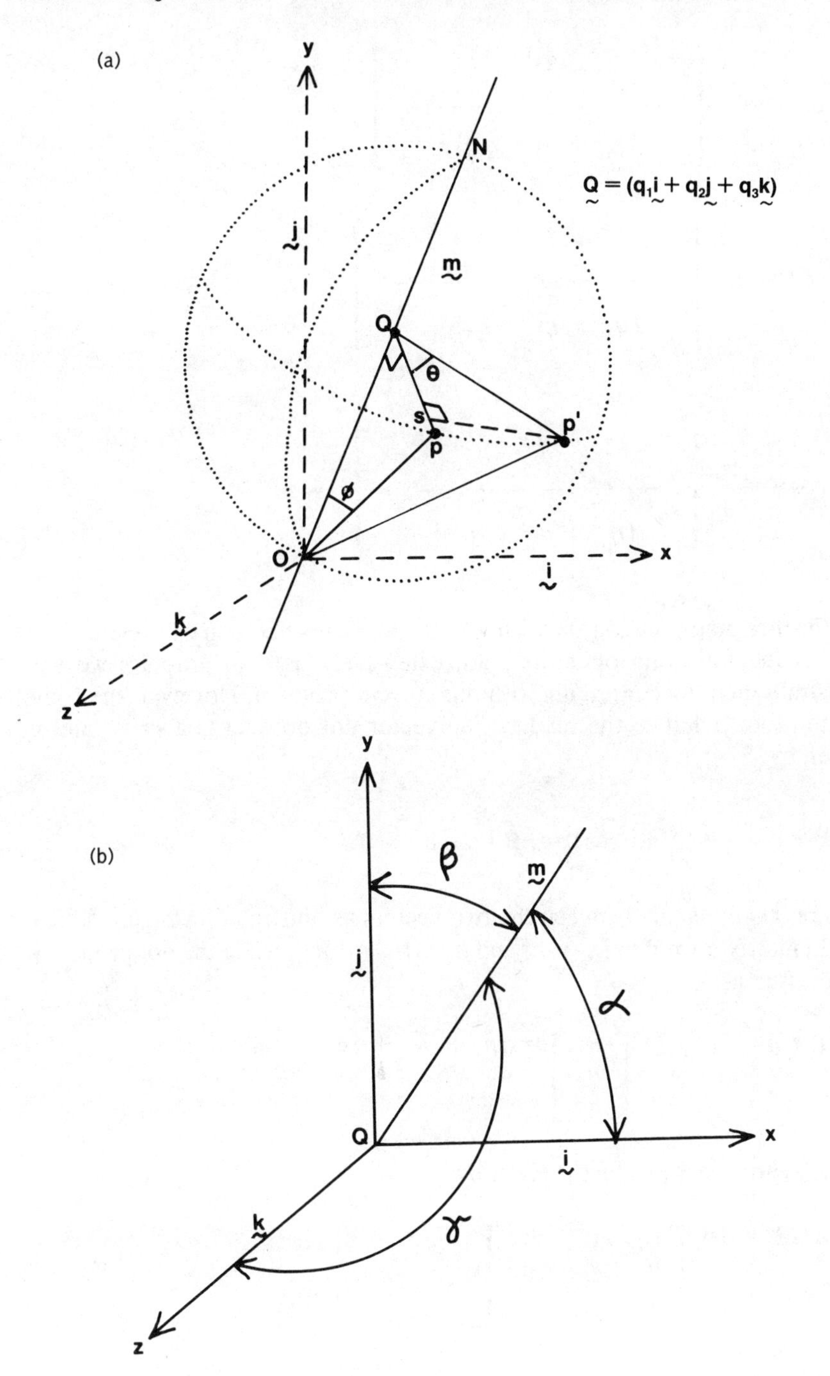

with **i, j, k** unit vectors in the x, y, z directions respectively. From the above we derive the following:

$$\cos\alpha = \left[\frac{q_1}{(q_1^2 + q_2^2 + q_3^2)^{1/2}}\right] = n_1$$

$$\cos\beta = \left[\frac{q_2}{(q_1^2 + q_2^2 + q_3^2)^{1/2}}\right] = n_2$$

$$\cos\gamma = \left[\frac{q_3}{(q_1^2 + q_2^2 + q_3^2)^{1/2}}\right] = n_3$$

Further manipulation through vector calculus is necessary to relate P' to P. As the following operations are extensively used in graphics we will illustrate them by continuing with our current problem. However, the complete proof is left to the reader. The vector dot product is a scalar and is given by:

$$\mathbf{P} \bullet \mathbf{n} = |\mathbf{P}|\,|\mathbf{n}| \cos\theta = |\mathbf{P}| \cos\theta$$

where θ is the angle between the two vectors as illustrated in Figure 5.3(a).

In matrix form $\mathbf{P} = [x \ \ y \ \ z]$ and $\mathbf{n} = [n_1 \ \ n_2 \ \ n_3]$, while the dot product is expressed as:

$$\mathbf{P} \bullet \mathbf{n} = [x \ \ y \ \ z] \begin{bmatrix} n_1 \\ n_2 \\ n_3 \end{bmatrix} = xn_1 + yn_2 + zn_3$$

The vector cross product is given by:

$$\mathbf{n} \times \mathbf{P} = \det \begin{bmatrix} \mathbf{i} & \mathbf{j} & \mathbf{k} \\ n_1 & n_2 & n_3 \\ x & y & z \end{bmatrix}$$

$$= \begin{vmatrix} n_2 & n_3 \\ y & z \end{vmatrix} - \begin{vmatrix} n_1 & n_3 \\ x & z \end{vmatrix} + \begin{vmatrix} n_1 & n_2 \\ x & y \end{vmatrix}$$

$$= \mathbf{i}(n_2 z - n_3 y) - \mathbf{j}(n_3 x - n_1 z) + \mathbf{k}(n_1 y - n_2 x)$$

The matrix form of the cross product is given by:

$$\mathbf{n} \times \mathbf{P} = [x \;\; y \;\; z] \begin{bmatrix} 0 & n_3 & -n_2 \\ -n_3 & 0 & n_1 \\ n_2 & -n_1 & 0 \end{bmatrix}$$

It also holds that

$$\mathbf{n} \times \mathbf{P} = |\mathbf{n}| \; |\mathbf{P}| \sin \varphi = |\mathbf{P}| \sin \varphi$$

We now state the following equation and omit the derivation.

$$\mathbf{P}' = (\mathbf{P} \bullet \mathbf{n}) \, \mathbf{n} \, (1 - \cos \theta) + \mathbf{P} \cos \theta + (\mathbf{n} \times \mathbf{P}) \sin \theta$$

The equation gives the transformed point in terms of:

(1) The initial point.
(2) The angle of rotation.
(3) The direction of the axis rotation.

In matrix form

$$\mathbf{P}' = [x \;\; y \;\; z] \begin{bmatrix} n_1 \\ n_2 \\ n_3 \end{bmatrix} [n_1 \;\; n_2 \;\; n_3] \, (1 - \cos \theta) + [x \;\; y \;\; z] \cos \theta$$

$$+ [x \;\; y \;\; z] \begin{bmatrix} 0 & n_3 & -n_2 \\ -n_3 & 0 & n_1 \\ n_2 & -n_1 & 0 \end{bmatrix}$$

and using homogenous coordinates

$$\mathbf{P}' = [x \;\; y \;\; z, 1] \left\{ \begin{bmatrix} n_1 n^2 & n_1 n_2 & n_1 n_3 & 0 \\ n_1 n_2{}^2 & n_2 & n_2 n_3 & 0 \\ n_1 n_3 & n_2 n_3 & n_3{}^2 & 0 \\ 0 & 0 & 0 & 1 \end{bmatrix} (1 - \cos \theta) \right.$$

$$+ \begin{bmatrix} 1 & 0 & 0 & 0 \\ 0 & 1 & 0 & 0 \\ 0 & 0 & 1 & 0 \\ 0 & 0 & 0 & 1 \end{bmatrix} \cos \theta$$

$$+\begin{bmatrix} 0 & n_3 & -n_2 & 0 \\ -n_3 & 0 & n_1 & 0 \\ n_2 & n_1 & 0 & 0 \\ 0 & 0 & 0 & 1 \end{bmatrix} \sin\theta$$

The three terms within the braces give the required rotation matrix [**R**]. They can be combined and written as in the beginning of the chapter.

5.9 PROJECTIONS

In the above section we discussed several coordinate systems that are available to the investigator, with the proviso that the ultimate one must be rectangular for graphics purposes. It is usually the case that the three-dimensional objects that we manipulate are defined in a three dimensional world coordinate system, but the display surface itself is two-dimensional. The problem is to manipulate this three-dimensional data in such a way that the resultant two-dimensional image 'looks' three-dimensional. The mathematical tools to carry out these conversions fall within the domain of projective geometry and the process itself is termed projection. There are several types of projection but we limit ourselves to the most common forms known as orthographic parallel and perspective. The prevalent types of orthographic projections are the front elevation, top plan and side elevation. In these three cases the projection plane is perpendicular to one of the principal axes.

In the following discussion we shall refer to the display surface as the view plane and invisible directed vectors as projectors. These projectors can be thought of as invisible beams of light radiating from a point. When the projectors intersect the view plane they become visible.

Given that we shall confine ourselves to parallel and perspective projections it is natural to ask what is the difference? The main difference between them lies in the angle of projectors as radiated from their origin to the image(s). In parallel projection, projectors proceed in parallel lines toward the object(s). When the relevant projectors intersect the view plane they become visible on the display surface. In this case the projectors are parallel because their origin is set at an infinite distance from the point at which the observer is located. This point is known as the view reference point.

In a perspective projection, the projectors radiate from a point set at a finite distance from the view reference point. As in the parallel case, they proceed until they touch the object. Where the projectors intersect the view plane they become visible on the display surface. These two forms of projection are illustrated in Figure 5.4. Another important aspect of these two projections is the effect of the projection on the size of the object(s). In parallel projection the size of the object is unaltered by the distance from the viewer. In other words moving the view reference point closer or further away has no effect on the resultant image. This is clearly because parallel lines do not converge in the distance as is the case with perspective projections. On the other hand, in perspective projections, size is a function of distance. That is, the closer an object is to the viewer, the larger it appears to be (Figure 5.5).

Figure 5.4: Parallel and Perspective Display of an Object in Three Dimensions

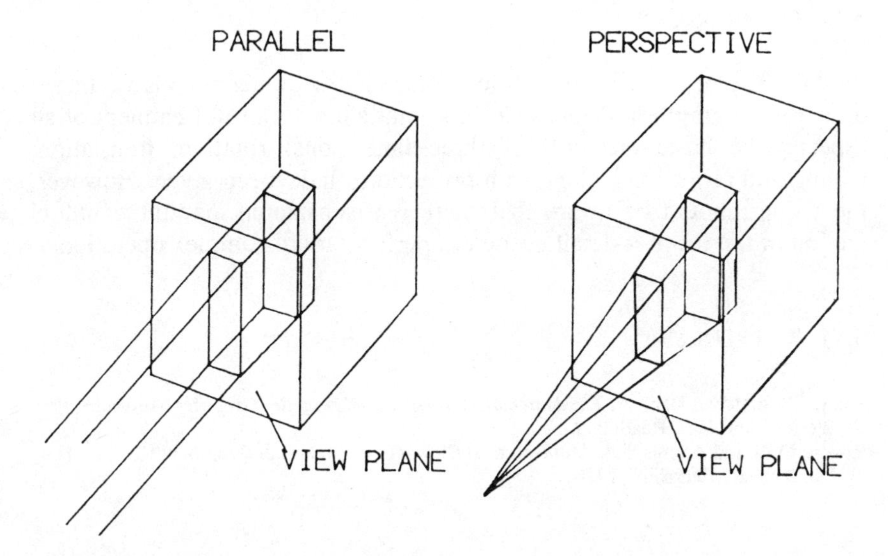

Figure 5.5: Perspective Projections of a Cube onto Two View Planes (Left) Between Observer and Object, and (Right) Behind the Object. The projection of point *p* onto both view planes is shown

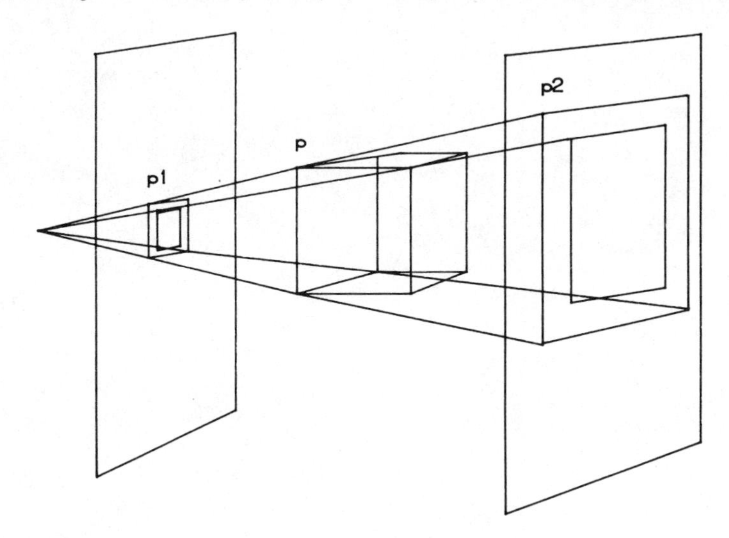

To achieve a point perspective (and without going into all the detail), we may refer to the 4 × 4 transformation matrix. The elements that control the perspective are the first three row elements in the last column. That is $a41$, $a42$, $a43$ in our previous notation. For a detailed description of projections the reader is referred to Foley and van Dam (1982) and for a more mathematical treatment to Rogers and Adams (1976).

5.10 CONCLUSIONS

In this chapter we have presented some of the fundamentals of three-dimensional graphics. While space does not allow us a full treatment of all aspects, the basic elements of three-dimensional rotation, translation, scaling and reflections, along with projections, have been given. However, the reader should be aware that there is a substantial amount of information in the form of detail needed to perform more complex operations.

5.11 REFERENCES

Foley, J.D. and Van Dam, A. *Fundamentals of Interactive Computer Graphics* (Addison-Wesley Publishers, Reading, Massachusetts, 1982)

Rogers, D.A. and Adams, J.A. *Mathematical Elements for Computer Graphics* (McGraw-Hill, Maidenhead, Berkshire, 1976)

6 Hidden Lines and Hidden Surfaces

6.1 AN INTRODUCTION TO HIDDEN LINES AND SURFACES

In general the objects that are manipulated in three-dimensional graphics can be classified into two groups: wire frame and true solids. In the wire frame group the objects are described in terms of sets of lines, and in the solid group the objects would be described in terms of surfaces with well-defined properties. While both groups of objects are quite distinct visually, they do share common problems: the removal of unwanted lines and/or surfaces respectively.

What do we mean by unwanted lines or surfaces? In the context of computer graphics this would mean unseen lines or surfaces if the nearest surfaces (as defined by sets of lines in the wire frame case) are considered opaque. Look at the two wire frame cubes illustrated in Figure 6.1. The cube in Figure 6.1(a) is the 'transparent' rendering, in that every line is visible from any given viewpoint. In Figure 6.1(b), however, the surfaces nearest the observer have been made opaque and the cube has a solid appearance. In effect, the process of defining opaque surfaces on wire frame objects consists of eliminating the lines that would be obstructed visually if the surfaces in front of them are opaque.

This elimination process can be quite simple or very complex, depending on the objects under consideration and the methods employed to describe those objects. Examples of hidden line elimination can be found in the serial programs described in Chapter 8, while hidden surface elimination techniques would be applicable to the solid molecular modelling systems described in Chapter 10.

Figure 6.1: Two Views of a Cube. (a) Wireframe Representation; (b) With Hidden Lines Removed

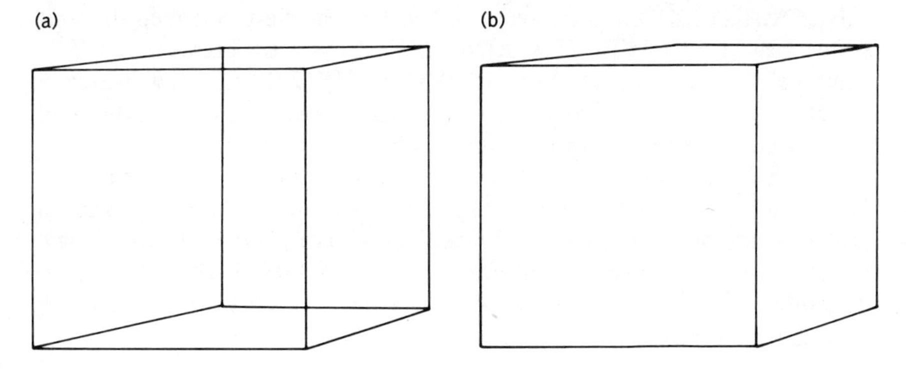

6.2 A SIMPLE HIDDEN LINES ALGORITHM

We will now consider a simple algorithm for the removal of hidden lines designed by Angell (1981). First, however, we must define in a very precise manner the types of objects that we are going to deal with and the nature of the data needed to accurately describe these objects. In this elementary example we shall only consider objects that consist of convex bodies, where a convex body in three dimensions is one in which a line between any two points in the body lies totally inside the body. A further restriction with this algorithm is that the origin with respect to the observer's position lies inside the body. These are quite restrictive conditions, especially the restriction to convex objects, and as such the algorithm is not of general use.

We must now determine the data required to precisely define the objects while at the same time allowing us to eliminate the hidden lines. A primary task of hidden line algorithms is to check if lines lie behind plane segments (facets) on the surface of the solid. As these are wire frame objects we would certainly require a set of points. The points can be grouped into pairs forming lines and the lines can be grouped into bounding facets (a description of simple array data structures for describing facets may be found in Chapter 7). There is always a finite amount of storage space available on any machine, no matter how large. Thus there must be limits on the amount of point, line and facet information that we can store. This can of course be quite large for any reasonable system. To expedite the search routines there is also a limit on the number of lines that can compose a bounding facet. In the case described by Angell (1981), the limit is six. However, objects with more than six lines defining a facet may be handled by flagging those facets and subjecting them to a subdivision process. Because of the above restrictions we know that any facet may be seen if and only if the infinite plane containing the facet cuts the line joining the eye (viewpoint) to the origin between these two points.

The object, is of course, to discover whether the Kth facet is visible or not. The Angell algorithm takes three points $((xi, yi, zi) 1 i{=}1,2,3)$ from the facet. Suppose that INDEX1 = LINEFACET(1,K) and INDEX2 = LINEFACET(2,K) are the indices of two lines in the facet; then the four vertex indices LINEVERTEX(1,INDEX1), LINEVERTEX(2,INDEX1), LINEVERTEX(1,INDEX2) and LINEVERTEX(2,INDEX2) will contain the indices of three non-collinear points. These three vertices can be used to calculate the equation of the plane containing the facet, $ax + by + cz = d$. In functional form the equation becomes: $f(x,y,z) = ax + by + cz - d$. From this we can discover if the origin (0,0,0) and eye (0,0,−DIST) lie on the same side of the plane by comparing the sign of $f(0,0,0)$. That is to say, we compare the sign of $-d$ with the sign of f(0,0,DIST). This amounts to determining the sign of $-d - c\,x$ DIST. The facet is visible if and only if $d + c\,x$ DIST is of opposite sign to d.

This calculation can be made for every facet; if the facet is visible, the indices of all its boundary edges are added to an array. Clearly, a line that is a boundary between two visible facets will appear twice in the array. To expedite the algorithm, the elements of the array are arranged in numerical order and the duplications are deleted. The resultant array contains the indices of all the visible lines in the object. Using this information and that contained in the two-dimensional array LINEVERTEX, all of the visible lines may be drawn to complete a hidden line perspective image. The complete details of this algorithm and a more general one, including the FORTRAN source code, are given in Angell (1981).

6.3 THE GALIMBERTI AND MONTANARI ALGORITHM

We now present another hidden lines algorithm developed by Galimberti and Montanari (1969). As we stated in the introduction to this chapter there are numerous algorithms which operate on various types of objects, or to be precise, operate on various descriptions of those object. In the Galimberti and Montanari algorithm, the objects can be concave and convex plane-faced. In this case, all of the edges of the objects are considered sequentially and all planes which hide every point of an edge are determined.

The algorithm may be summarised as follows:

Step 1. Every edge is recognised as corresponding to a convex or concave dihedral. A dihedral angle is the angle formed by two intersecting planes. The dihedral angle is measured in a plane perpendicular to the line of intersection of the two planes. For an illustration of this see Figure 6.2.

Step 2. All lines hidden by their own volume are eliminated; here it is necessary to know whether an edge is concave or not. At this point, if it is found that we have to represent a convex object only, the program will automatically terminate.

Step 3. All the edges not hidden by their own volume are sequentially examined. For each edge, one of the two extreme points is considered. The primary task at this point is to determine the set of faces that hide this point, that is the 'nature' of this point. Then, starting from the chosen point, the segment is examined and the set is modified by making use of intersections of the projected segments on the picture plane. Clearly, at the end of this step, the 'nature' of the other extreme point of the segment is known. Thus, the nature of only a few points need to be determined.

At this point it might be instructive to compare the information required in the Angell and Galimberti algorithms. In the case of the latter, objects in three-dimensional space are considered to be bounded by faces, which are

Figure 6.2: A Dihedral Angle is Formed by Two Intersecting Planes. The angle (theta) is measured in a plane perpendicular to the line of intersection of the two planes (*M* and *N* here)

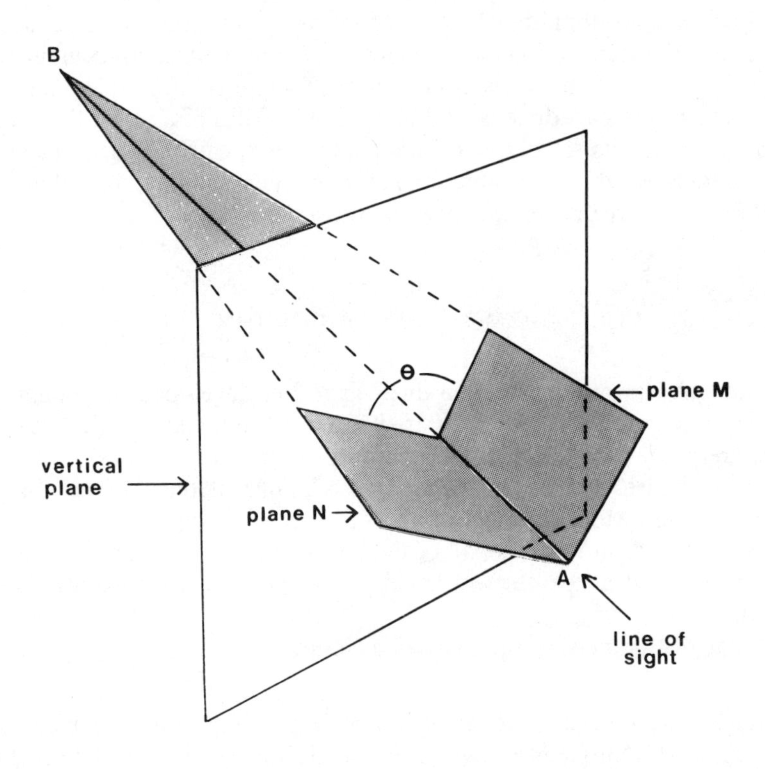

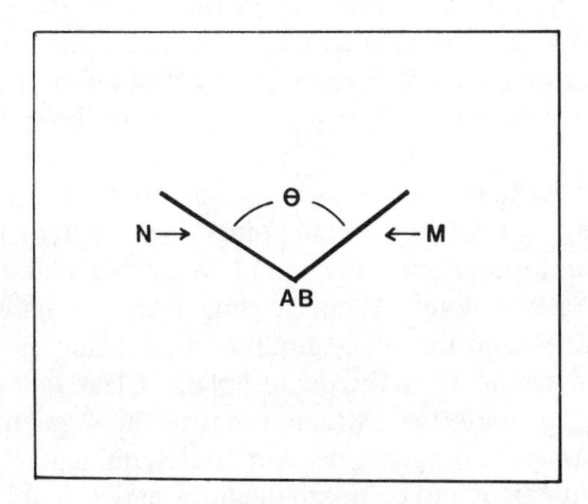

portions of a plane. The faces are always considered closed — that is, enclosing their boundary. Each face is bounded by edges and the two extreme points of an edge are nodes (the reader should note that nodes and vertices are identical). Every edge belongs to two faces. An edge is convex (concave) if it corresponds to a convex (concave) dihedral. A node is concave if it belongs to at least one concave edge; otherwise, it is convex.

It would appear that a greater variety of information is needed for the Galimberti algorithm than is required for the Angell case. And, to some extent this is true. This is of course where the trade-offs start appearing with regard to the algorithms and the run times. Also recall that the Angell algorithm could only be used on single convex bodies, whereas the Galimberti algorithm has no such restriction.

6.4 THE HIDDEN SURFACE PROBLEM

A hidden surface algorithm consists of several major components. From a functional point of view, the hidden surface algorithm may be defined as a quintuple (a function with five variables) (Giloi, 1978). While this approach may seem a little formal, it does have the advantage of clearly specifying the parameters that are required. According to Giloi, the components of this quintuple are: a set of objects in three-dimensional space; a set of visible segments (that is, a projection of the three-dimensional objects) in two-dimensional space; a set of 'intermediate representations' (the state of the objects during their transformation from three-dimensional to two-dimensional); a set of 'transition functions'; and a 'strategy function' specifying the order in which these functions are to be applied.

This set of transition functions consists of the following six functions:

(1) A function that produces a perspective view — that is to say, a projective mapping; thus, the domain of the function is three-space and the range is two-space.

(2) A function that calculates the point of intersection of two line segments; in this case the domain and range are the same, that is, two-space or three-space.

(3) A function that performs a 'containment test' in two-space; the containment test would check whether or not a point is in a given bounded surface. In the Giloi definition, the result of this test is a boolean variable which is true if the point is contained, and false otherwise.

(4) A function that performs a 'depth test'; in essence, this test compares two points and determines which has the greater depth. In the computer graphics context, the greater the depth the farther away from the point of observation (the eye position).

(5) A function that performs a 'visibility test' for a given surface; in the Giloi definition the result of this function is a boolean value which is true if the surface is potentially visible and false if the surface is totally invisible.

6.5 A PRELIMINARY CLASSIFICATION

To assist us in our understanding of the objects that are considered in some hidden surface algorithms, we now consider a preliminary classification with respect to the degree of object reduction (Giloi, 1978). This classification will take the form of a rooted tree structure. At the root is the problem of hidden surface algorithms for opaque objects. As this is a reduction classification we next consider the nature of the faces of the objects. In general we can define two distinct sets of faces: on the one hand we have opaque objects with plane faces and on the other we will have opaque objects with curved faces. A precise description of the mathematics of plane and curved faces may be found in Rogers and Adams (1976).

Let us consider the plane faces first. This set can be partitioned into two classes: in the first class the most complex entity is a solid (that is, a face with depth information). Objects are rendered as a set of contour lines and lines of intersection; in the second class the most complex entity is a polygon (faces, without depth information). Objects in this class are also rendered as a set of contour lines and lines of intersection or as shaded objects.

We next consider the curved faces. As in the plane face case we can partition the curved faces into two classes: in the first class the most complex entity is a surface, given as a grid of points — objects are rendered as a net of grid lines; in the second class the objects are defined analytically and rendered as a set of contour lines and lines of intersection.

6.6 SURFACE REPRESENTATION AND HIDDEN SURFACE METHODS

Before discussing several methods of hidden surface identification and elimination, we look at the problem of polygon infill — that is, the process of determining the boundary of a closed polygon and filling this enclosed region with some feature such as colour. There are several methods available for this process and a complete treatment is given in Foley and Van Dam (1982).

The method that we have chosen to illustrate this process is known as polygon scan conversion. At this point we must explicitly state the difference between region filling and polygon scan conversion. In the region filling process, the region is defined by pixel values in the refresh (frame) buffer. In polygon scan conversion, the region is defined by the polygon vertices and absolutely no assumptions are made with respect to the contents of the refresh buffer.

In Figure 6.3 we illustrate a basic polygon scan conversion process, with a single scan line passing through the polygon. The object of the process is to determine which pixels on the scan line are within the polygon, and set these pixels to their corresponding values. For the polygon and scan line illustrated in Figure 6.3, we see that pixels with *x* coordinates 2 through 4 and 8 through 12 lie on the scan line. This process is, of course, repeated for each scan line which intersects the polygon. In general the process would require three steps per scan line. These steps may be described as follows:

(1) Determine the intersections of the scan line with all edges of the polygon.
(2) Sort the intersections by increasing *x* coordinate.
(3) Fill in all pixel pairs of intersections.

An obvious method to use in deciding which pixels on the scan line are

Figure 6.3: Infilling a Polygon Using Scan Line Conversion. The polygon is defined by edges *e*1 to *e*6. Numerals refer to *x*,*y* coordinates of pixels (shaded)

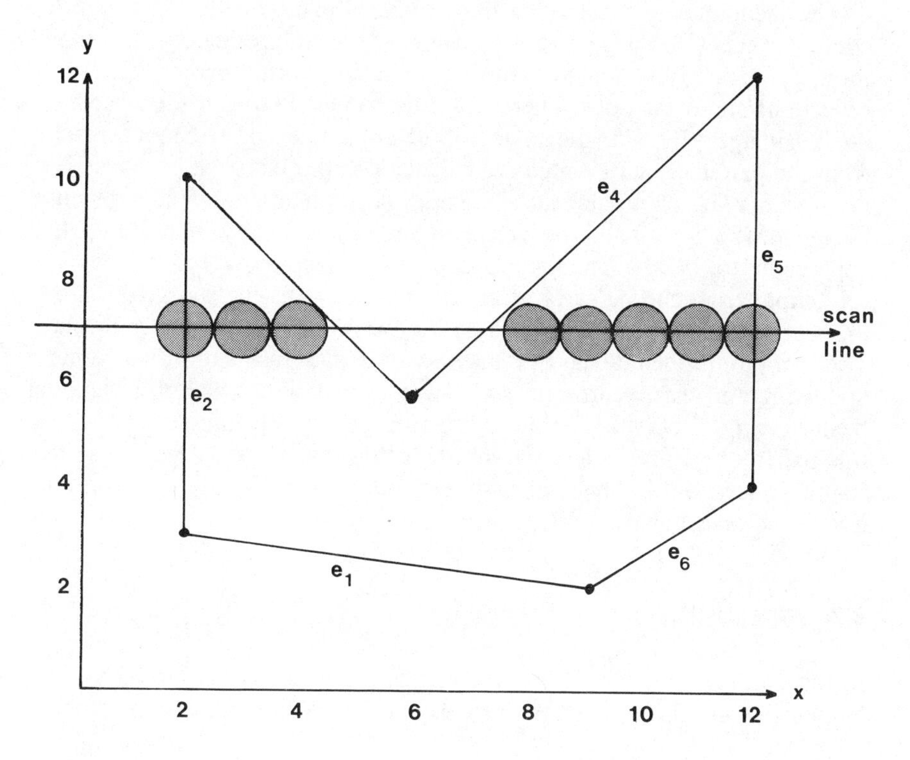

in the polygon is to test each and every pixel. Clearly, this is rather inefficient, in that in general a sequence of adjacent pixels will lie within the polygon. This observation is based on the notion of spatial coherence. That is, it is often the case that the polygon does not change as we move from pixel to pixel or from scan line to scan line. Thus, the search is reduced to those pixels at which changes occur.

Ray tracing and scan line algorithms are two of the most widely used methods for creating frame buffer images of three-dimensional objects. Ray tracing generates an image by casting a ray of light from the eye point through each pixel of the image and into the scene. The visible surfaces are determined by testing for the line–surface intersections between the ray and each object in the scene. Considerable realism can be imparted to the image by recursively tracing reflected and refracted rays. The primary advantages of the ray tracing methods are the ability to use a global lighting model that calculates reflections, refractions and shadows and the facility to handle a variety of geometrical primitives.

By contrast, a scan line algorithm generally takes advantage of coherence to draw the surfaces incrementally. The main advantage of the scan line method is that the incremental calculation of geometry is very efficient. On the other hand, the primary disadvantage is that the local lighting method is not as realistic as the global model.

The technique of ray tracing has produced some of the most realistic computer generated images to date. Some of the features incorporated into these systems would include: surface texturing, local shading, shadows, reflections and refractions. There is a price to pay, however, for these high quality images. Indeed, the major disadvantage of this technique results from the extremely CPU-intensive calculations that must be performed as ray tracing relies on a point sampling approach. Other disadvantages would be the possibility of aliasing artefacts and the difficulty in using spatial coherence to best advantage, because the shapes of reflections and refractions from curved surfaces are so complex. Aliasing is a defect in the image created by an improper sampling procedure on the raster display. The symptoms include jagged lines, Moire patterns, and small objects appearing and disappearing in successive frames. For a superb article on beam tracing of polygonal objects see Heckbert and Hanrahan (1984). In this paper, the authors describe an algorithm that does indeed utilise the spatial coherence of polygonal environments by combining features of both image and object space hidden surface algorithms.

6.7 CONCLUSIONS

In this chapter we have discussed some of the issues of three-dimensional hidden line and hidden surface representations. While hidden line elimination is in general easier to carry out than hidden surface removal, they can both (depending upon the objects) be extremely difficult. In the ordinary activities of the biologist, we suspect that the elimination of hidden lines would be of more use than surface removal. However, if you are planning to produce very realistic images of, say, three-dimensional molecules, then you will necessarily be faced with the problems of hidden surface removal and the most suitable algorithm will have to be chosen.

6.8 REFERENCES AND BIBLIOGRAPHY

Angell, I.O. *A Practical Guide to Computer Graphics* (Macmillan Press, London, 1981)

Avis, D. and Toussaint, G. 'An Optimal Algorithm for Determining the Visibility of a Polygon from an Edge', *IEEE Transactions on Computers*, C-30 (12) (1981), 910–14

Blinn, J.F. *et al.* 'Scan Line Methods for Displaying Parametrically Defined Surfaces', *Communications of the ACM*, 23 (1) (1980), 23–4

— and Newell, M.E. 'Texture and Reflection in Computer Generated Images', *Communications of the ACM*, 19 (10) (1976), 542–7

Boissonnat, J.D. 'Geometric Structures for Three-dimensional Shape Representation', *ACM Transactions on Graphics*, 3 (4) (1984), 266–86

Crow, F.C. 'Shadow Algorithms for Computer Graphics', *Computer Graphics*, 11 (2) (1977), 242–8

Franklin, W.R. 'A Linear Time Exact Hidden Surface Algorithm', *Computer Graphics (ACM SIGGRAPH)*, 14, (3) (1980), 117–123

Foley, J.D. and Van Dam, A. *Fundamentals of Interactive Computer Graphics* (Addison-Wesley, Reading, Massachusetts, 1982)

Galimberti, R. and Montanari, U. (1969) 'An Algorithm for Hidden Line Elimination', *Communications of the ACM*, 12 (4) (1969)

Giloi, W.K. *Interactive Computer Graphics: Data Structures, Algorithms, Languages* (Prentice-Hall, Englewood Cliffs, New Jersey, 1978)

Heckbert, P.S. and Hanrahan, P. 'Beam Tracing Polygonal Objects', *Computer Graphics*, 18 (3) (1984)

Requicha, A.A.G. and Voelcker, H.B. 'Solid Modelling: A Historical Summary and Contemporary Assessment', *IEEE Computer Graphics Applications*, 2 (2) (1982), 9–24

Rogers, D.A. and Adams, J.A. *Mathematical Elements for Computer Graphics* (McGraw-Hill, Maidenhead, Berkshire, 1976)

Sutherland, I.E., Sproull, R.F. and Schumacher, R.A. 'A Characterization of Ten Hidden Surface Algorithms', *Computing Surveys*, 6 (1) (1974) 1–55

Weiler, K. and Atherton, P. 'Hidden Surface Removal Using Polygon Area Sorting', *Computer Graphics*, 11 (2) (1977), 214–22

Whitted, T. 'An Improved Illumination Model for Shaded Display', Communications of the ACM, 23 (6) (1980) 343–9

7 Graphical Representation of Biological Data

7.1 INTRODUCTION

We have looked at the manipulation of two and three-dimensional data in the last three chapters, concentrating mainly on general methodologies. We will now turn to consider specific biological applications of computer graphics. The present chapter deals with aspects of simple data manipulation, while Chapters 8 to 11 are concerned with areas of specific applications.

Three main areas have been chosen for discussion here. Firstly, we look at graphs and histograms. These mandatory weapons in the biologist's armoury have already been mentioned in Chapter 1, and commercial software for drawing graphs was also discussed in Chapter 3. The second section deals with other 'point plotting' techniques including map distributions and applications of point data transformations. We conclude this chapter by introducing some elementary graphics data structures, including a method for storage and recall of two and three-dimensional data for drawing objects.

7.2 GRAPHS AND HISTOGRAMS

Section 1.4 described the basic elements needed to draw a graph on a graphics device. You may recall that these elements comprised the axes, text, labels and data points. We also saw in Chapter 3 that graphics libraries are available to aid the mainframe and minicomputer user to draw graphs (we specifically discussed the SIMPLEPLOT and PLOTALL packages). Such packages take much of the 'donkey work' out of graph creation so that factors like axis drawing, text placement and scaling of the data are handled by the package being used. The user still has to learn a set of instructions to input the graph labels and parameters, but the chore of deciding on the positions of the graph elements 'on screen' is performed automatically.

Many computer users do not have access to a graphing package, or may wish to control the layout of a graph on a more 'personal' basis than is allowed by an available graph package. If such is the case, a first step will probably be to write a simple graphing program of the type outlined in Chapter 1. The following elementary program in Basic allows plotting of graph data. We have used Basic for the example here partly because of the comprehensive text manipulation commands inherent in standard Basic. Use of Basic also makes the programs implementable on any microcomputer with a Basic interpreter and graphics capability, and micro users probably have less access to graphing packages than large computer users.

The graphics commands in this Amstrad Basic program will need to be converted in order to run the program on other microcomputers. These commands are listed below.

Line	Command
40,50	INK sets up drawing colours
80	MODE sets 80 column text mode
200,250	MOVE moves pen to point (x,y)
210,220	DRAW draws a line to point (x,y)
340	LOCATE start plotting text at column,row
410	TAG change text plot to graphics mode (that is, text plotted at (x,y) coordinates instead of rows/columns).

The Amstrad computers use a 640 × 400 unit graphics screen, so the x and y scales will need to be adjusted for microcomputers with differing resolutions. Note also that the origin (that is, 0,0) is at the lower left corner of the screen.

The GRAPH program plots data points, but it is easy to amend the program to join up the data points. Both kinds of output are shown in Figure 7.1 and program 7.1.

Figure 7.1: A Graph Drawn Using the GRAPH Program Listed in Program 7.1. (a) Plotting of individual points; (b) Graph made up of joined lines

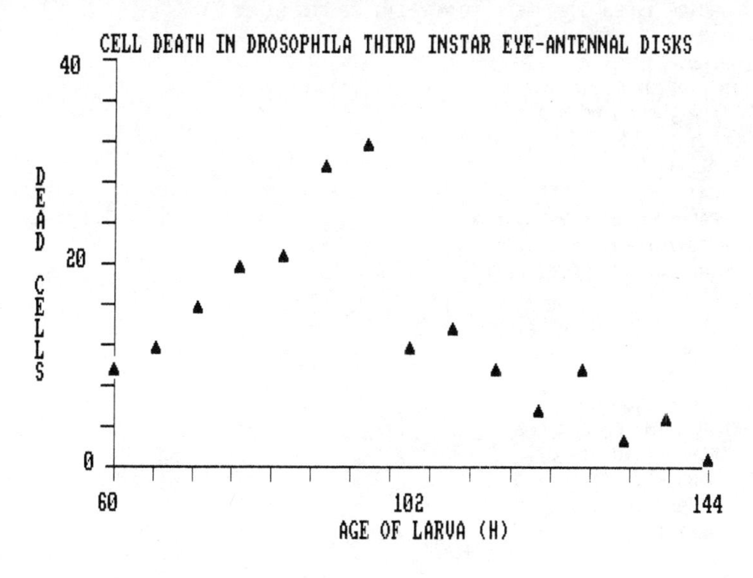

Figure 7.1 continued

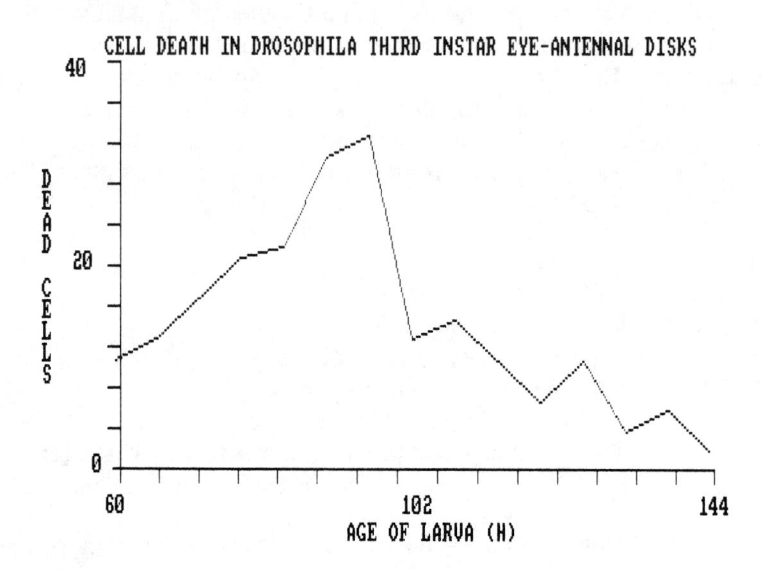

Program 7.1

```
10 REM ****PROGRAM SUPERG****
20 REM VERSION OF GRAPH PROGRAM TO PLOT TWO VARIABLES
30 REM TO DRAW A SIMPLE LABELLED GRAPH
40    INK 0,13
50    INK 1,0
60    READ TITLE$
70    CHOICE=1:REM USE 1 FOR POINT PLOT, 2 FOR LINES
80 MODE 2
90    READ POINTS
100 DIM X(POINTS),Y(POINTS)
110    FOR I=1 TO POINTS
120 READ X(I)
130 READ Y(I)
140    NEXT I
150    READ XMIN,XMAX,YMIN,YMAX
160    READ X$:REM X AXIS NAME
170    READ Y$:REM Y AXIS NAME
180 CLS
190 REM NOW DRAW THE AXES
200       MOVE 100,380
210       DRAW 100,80
220       DRAW 550,80
230 REM PUT IN SCALE MARKS
240    FOR I=1 TO 11
250       MOVE 90,(I*30)+50
260       DRAW 100,(I*30)+50
270    NEXT I
```

```
280    FOR I=1 TO 16
290      MOVE (I*30)+70,70
300       DRAW (I*30)+70,80
310    NEXT I
320 REM PRINT TITLE
330    P1=LEN(TITLE$):P1=40-(P1/2)
340    LOCATE P1,1:PRINT TITLE$;
350 REM NOW LABEL AXES
360 REM POSITION X LABEL FIRST
370 REM START POSITION IS CENTRE PT ON X AXIS MINUS HALF STRING LENGTH
380    AX=(320-((LEN(X$)*8)/2))
390 REM START POSITION IS CENTRE PT ON Y AXIS PLUS HALF STRING LENGTH
400    AY=(220+((LEN(Y$)*16)/2))
410 TAG
420       MOVE AX,40
430    PRINT X$;
440 REM NOW PRINT Y LABEL VERTICALLY
450    FOR I=1 TO LEN(Y$):M1$=MID$(Y$,I,1)
460      MOVE 40,AY-((I-1)*16)
470    PRINT M1$;
480    NEXT I
490    MOVE 530,60:PRINT XMAX;
500    MOVE 50,382:PRINT YMAX;
510    MOVE 70,90:PRINT YMIN;
520    MOVE 80,60:PRINT XMIN;
530   MOVE 55,240:PRINT INT((YMAX+YMIN)/2);
540   MOVE 305,60:PRINT INT((XMAX+XMIN)/2);
550 REM NOW PLOT POINTS
560   IF CHOICE=2 THEN GOTO 640
570 FOR I=1 TO POINTS
580  XTOP=XMAX-XMIN:YTOP=YMAX-YMIN
590  XTRUE=XTOP-(XMAX-X(I)):YTRUE=YTOP-(YMAX-Y(I))
600    MOVE 96+(450*(XTRUE/XTOP)),86+(300*(YTRUE/YTOP))
610    PRINT CHR$(244);
620 NEXT I
630 |COPY:END
640 REM LINE PLOT SECTION
650 FOR I=1 TO POINTS
660  XTOP=XMAX-XMIN:YTOP=YMAX-YMIN
670  XTRUE=XTOP-(XMAX-X(I)):YTRUE=YTOP-(YMAX-Y(I))
680 IF I=1 THEN MOVE 96+(450*(XTRUE/XTOP)),86+(300*(YTRUE/YTOP))
690   DRAW 96+(450*(XTRUE/XTOP)),86+(300*(YTRUE/YTOP))
700 NEXT I
710 |COPY:END
720 REM DATA - IN SEQUENCE -  TITLE,NUMBER OF POINTS,XVAL YVAL FOR EACH POINT
730 REM                       XMIN, XMAX,YMIN,YMAX
740 REM                       X$,Y$
750 DATA "CELL DEATH IN DROSOPHILA THIRD INSTAR EYE-ANTENNAL DISKS"
755 DATA 8
760 DATA 60,10,72,16,84,21,96,32,108,14,120,6,132,3,144,1
770 DATA 60,144,0,40
780 DATA "AGE OF LARVA (H)"
790 DATA "DEAD CELLS"
```

Although histograms tend to be used more in business than scientific circles, they are of some interest to biologists and it is possible to extend the GRAPH program to draw histograms as well. Instead of plotting points, a rectangle is drawn at each point location, and the modification shown in Figure 7.2 will allow the graph program to draw a histogram.

The code to draw the histogram box looks like this:

```
MOVE  XP,YP
DRAW  XP−(D/2),YP
DRAW  XP−(D/2),YAX
DRAW  XP+(D/2),YAX
DRAW  XP+(D/2),YP
DRAW  XP,YP
```

where the actual data point is (XP,YP), D is the width of the box, and the Y value of the X axis is YAX.

It is often useful to be able to plot data in three dimensions, for example if concentrations are to be plotted over a two-dimensional area, or if two variables are to be plotted over a series of time increments. In such cases, perspective projection of the data can be used to produce a more pleasing image, and hidden line elimination will reduce overlapping effects. The

Figure 7.2: Modification Required to Use the GRAPH Program to Draw Histogram Boxes Instead of Points or Lines

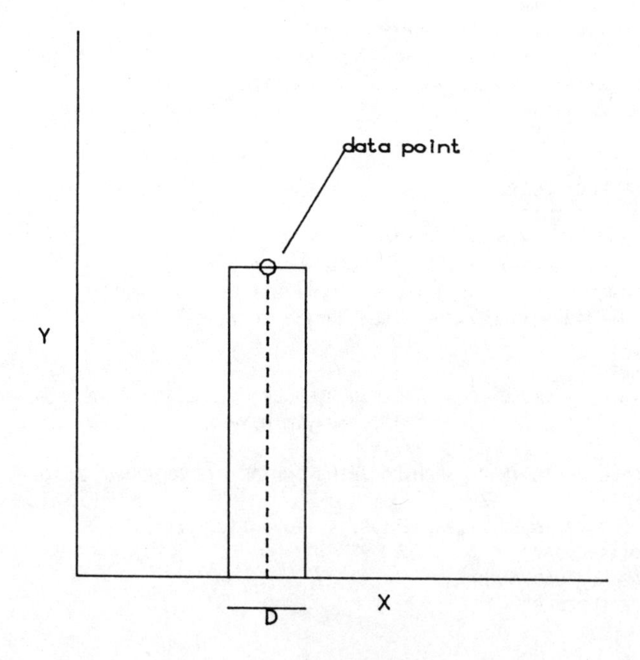

following elementary hidden lines algorithm can be used, if we assume that a stack of planes on the *Z* axis are to be plotted.

The first Z plane (that is, the one closest to the observer) is drawn, and variables YMIN and YMAX are set for each X increment. At this stage YMIN=YMAX, as only one value of Y has been plotted for each X increment. Each successive Z plane is processed as follows. For each possible X increment, the Y value is tested against the previous values of YMIN and YMAX for the X value. If the Y value is outside these bounds, the point is plotted, and YMIN or YMAX is reset to the new Y value, depending if it is above YMAX or below YMIN. If it is inside the bounds already set, the point is skipped.

Figure 7.3 shows the operation of the algorithm. A simple orthographic projection is illustrated, with the observer viewpoint along the Z axis. As the hidden lines removal works on the screen image after projection has been performed, it is independent of the projection method.

7.3 POINT PLOTS AND TRANSFORMS

A graph or histogram is often a convenient way of displaying correlation between variables, but many types of biological data do not fit into the

Figure 7.3: Plotting three-dimensional Data Using a Simple Hidden Lines Method. Dotted lines show hidden regions. Z planes are processed in the order 1..2..3.., and A..B..C.. represent the new values of YMAX for each section. YMIN will not be updated for the data shown here

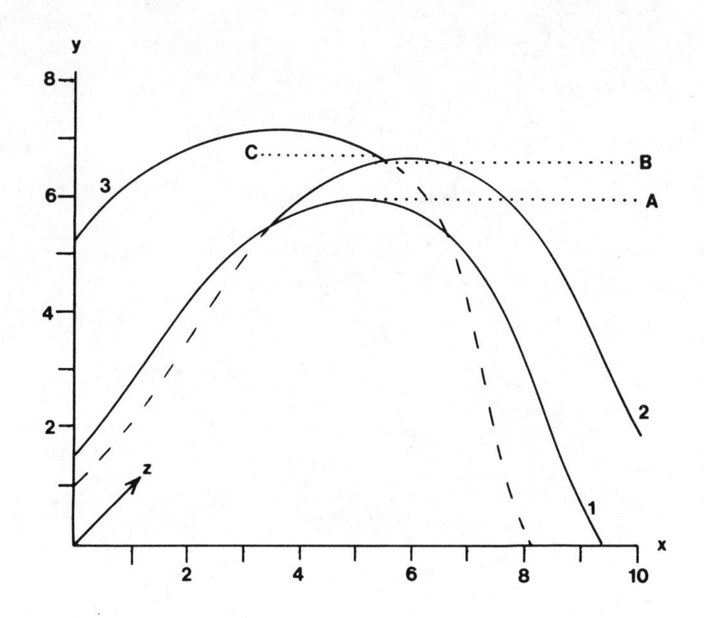

graph format. Exceptions of this type are often data distributions over a physical area, the space being perhaps a geographical region or part of an organism.

Figure 7.4 shows a typical geographical distribution of a bird population over the British Isles. The technique for plotting points on a distribution of this type is essentially the same as for plotting points on a two-dimensional graph: the same graphic commands are used. The difficulty lies in matching up positions of the points on the map, and this would be a tedious job if done by hand! Fortunately, the task is vastly simplified by use of a digitizer (discussed in Chapter 2). The relevant map can be placed on a digitizer surface and the outline may be traced using a hand cursor. The digitized outline can then be stored in the computer for future reference. Distribution points can be sent to the computer as required.

Figure 7.4: Geographical Distribution of Nightjars in the United Kingdom (Data from Gribble, 1984). Map and data digitized using a Calcomp 9000 series digitizer, output generated on a Tektronix 4663 plotter

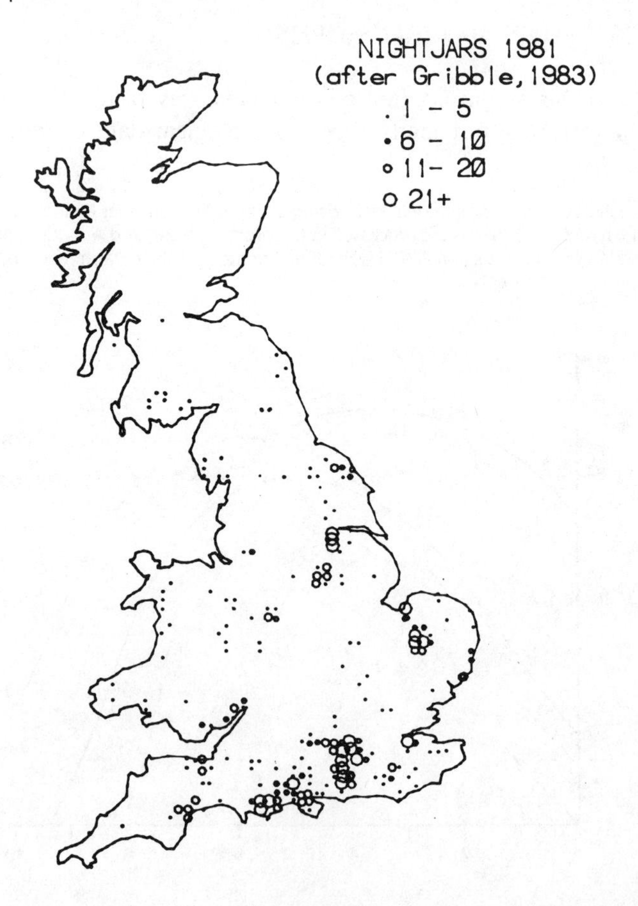

If a simple transformation is written into the digitizer program used to input the coordinates of the map or data points, the position of the map on the digitizer can be varied without rendering the data useless. Consider a micrograph on the digitizer surface (Figure 7.5). The coordinates sent to the computer from the digitizer will be x and y displacements from the digitizer origin. If the map 'corners' are sent to the computer before any other points are digitized, the x and y displacements of the corners from the origin can be calculated, and these displacements can be subtracted from the input data to centre the data around the origin. If the 'corner' data is

corner	x	y
top left	200	600
top right	700	600
bottom left	200	50
bottom right	700	50

the x and y displacements will be

$$(\text{XMAX} + \text{XMIN})/2$$

$$(\text{YMAX} + \text{YMIN})/2$$

so each x coordinate will have 450 subtracted from it, each y coordinate will have 325 subtracted from it.

The same digitizer program may be used to input other kinds of spatial

Figure 7.5: Digitization of Arbitrarily Positioned Data on a Digitizer Surface; *a, b, c, d* Are the Corner Reference Points for the Data

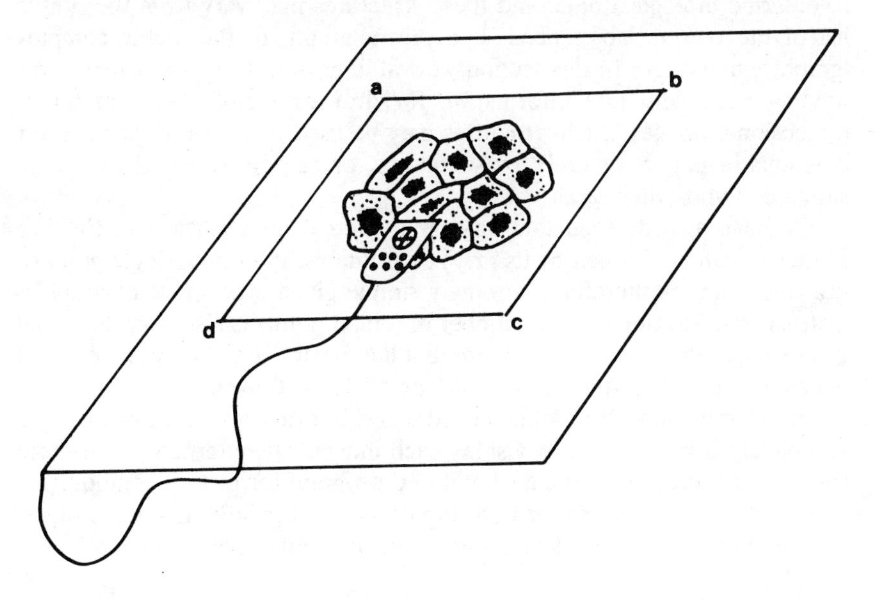

data, and it is often necessary to perform additional transformations on the data after the data has been stored in the computer. An example is the analysis of *Drosophila* (fruitfly) denticle belts. Let us first review the biological background. *Drosophila* has a total of 14 embryonic and larval segments, each with its own characteristic arrangement of hairs called denticles. The 'fingerprint' of these denticles has been used to discern which segments are present in larvae with mutant segments, and a computer pattern matching method has been used to perform the comparisons.

This study (by M. Carratala and R. Ransom, unpublished observations) involved three separate steps. In the first step, a digitizer was used to input denticle positions from photographs. The photographic material was carefully chosen to be of similar aged larvae and at the same magnification. The second step was transformation of the data to compensate for differences in larval size. Maximum and minimum x and y coordinates for each denticle band were calculated, and 'standard' size (600×150) distributions were calculated and were scaled to a common coordinate area (Figure 7.6). Finally, pattern recognition algorithms were devised and were applied to the transformed data. Program 7.2 shows the method for display of the data in Figure 7.6.

7.4 GRAPHICS DATA STRUCTURES

The study of graphics data structures is deserving of far greater attention than we can give in these few pages, and the interested reader may like to consult Giloi's book (1978). Many different data structures have been used to describe biological data, and these structures may vary from the simple list of the two variables needed to plot a graph, to the highly complex 'graphics metafiles'. In this section we will limit ourselves to a structure for holding point and line information for drawing simple two and three-dimensional images. Such structures may be used to construct pictures for diagrams in papers or OHP presentations, or can be extended for use in simulation and other areas.

We have already seen that any point on a display surface — CRT or plotter — can be defined by its (x,y) coordinates. Plotting a single point or drawing a line is therefore extremely simple given a basic set of graphics instructions. Plotting a small number of joined points is also easy, but what if we wish to draw a complex figure that has 50 or even 100 lines, not all of which are joined in one sequential length? Each drawing instruction in a program could specify absolute coordinates, but this would mean writing a completely new program to display each image. The alternative is to hold the data in a file and to use a generalised program for picture creation.

A standard procedure for both inputting and for holding data is therefore necessary, and the relevant data structures must specify three things:

Figure 7.6: Transformation of *Drosophila* Denticle Band Data. The two plots at the top of the figure show the digitized 'raw data' from two different segments. The bottom two plots show the transformed data from both segments transformed to fill the 600 × 150 unit area of the enclosing box. The transformed data can then be used as input to a pattern recognition program for comparative purposes

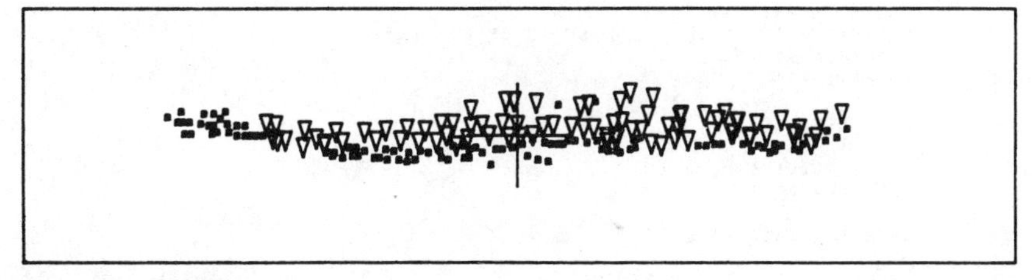

FILE = WTØ3A1

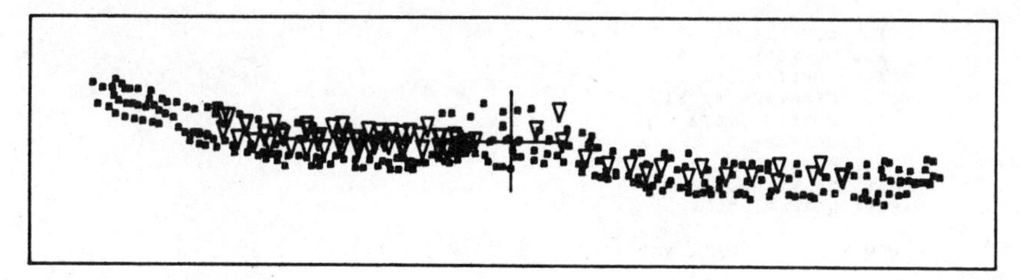

FILE = ABØ1A1

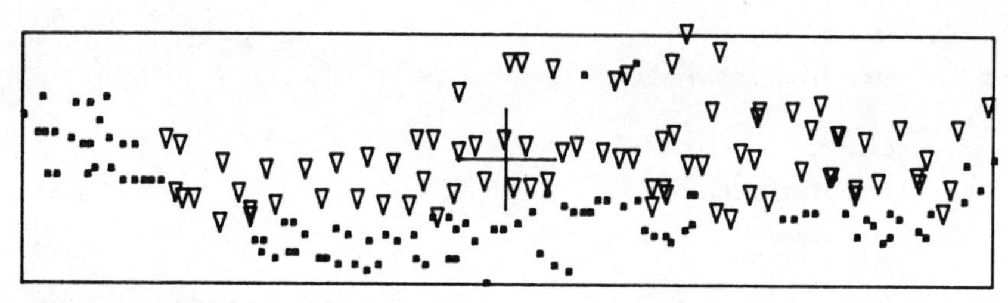

FILE = WTØ3A1T

FILE = ABØ1A1T

Program 7.2

```
C       THIS IS A FORTRAN PROGRAM USING THE TEKTRONIX PLOT-10 IGL
C       LIBRARY TO DISPLAY THE DATA OBTAINED FROM DIGITIZED
C       DROSOPHILA LARVAL SEGMENTS.  THE DATA IS HELD IN A
C       FILE AS A SERIES OF X,Y COORDINATES FOR EACH DENTICLE,
C       TOGETHER WITH ITS SIZE CLASS
C
        CHARACTER*12 FILE1
        WRITE(5,*)'TYPE IN THE FILE TO BE DISPLAYED'
        READ(5,101)FILE1
101     FORMAT(12A)
C
C       NOW SETUP IGL
C
        CALL IGLINI (ID,IB,IO)
        CALL NEWPAG
C
C       OPEN THE DATA FILE
C
        OPEN(4,NAME=FILE1,TYPE='OLD')
C
C       DRAW THE OUTLINE OF THE 600,150 ARRAY USED TO DISPLAY THE POINTS
C       TOGETHER WITH CENTRAL CROSS
C
        CALL MOVE(45.,62.5)
        CALL DRAW(55.,62.5)
        CALL MOVE(50.,67.5)
        CALL DRAW(50.,57.5)
        CALL MOVE(0.,50.)
        CALL DRAW(0.,75.)
        CALL DRAW(100.,75.)
        CALL DRAW(100.,50.)
        CALL DRAW(0.,50.)
C
C       NOW READ THE DATA AND PLOT IT WITHIN THE DISPLAYED OUTLINE
C
        DO 100 I=1,2000
        READ(4,*)X,Y,ISIZE
C
C       CHECK IF ALL DATA READ IN
C
        IF(X.EQ.9999)THEN
C
C       PLOT TITLE AT LOWER LEFT
C
        CALL MOVE (10.,30.)
        CALL TEXT(6,'FILE =')
        CALL MOVE (24.,30.)
        CALL TEXT(10,FILE1)
C
        CALL GRSTOP
        READ(5,*)DUMMY
        STOP
        END IF
C
C       ADJUST INPUT COORDINATES TO FIT IN DISPLAY AREA
C
        X=X/6
        Y=(Y/6)+50
C
C       CHOOSE AND DISPLAY MARKER VALUE FOR THIS DENTICLE
C       DEPENDING ON ITS SIZE CLASS
C
        IMARK=ISIZE+9
        IF(IMARK.EQ.11)IMARK=10
        IF(ISIZE+9.EQ.10)IMARK=0
        CALL MARKER(X,Y,IMARK)
C
100     CONTINUE
200     CONTINUE
C
        STOP
        END
```

(1) The (x,y) locations of all the points in the figure.
(2) The ordering of the points (that is, the order in which they are to be plotted).
(3) The connection between the points (are two consecutive points to be joined or not?)

A simple way of representing the data is as follows: we first take two one-dimensional arrays X(1 ... *n*), Y(1 ... *n*), where *n* is the total number of points to be drawn. Of course, both X and Y must be dimensioned the same, as each point has both *x* and *y* coordinates. We will call these arrays the coordinate data. As X(1) precedes X(2), these arrays also allow the ordering of the data: point X(1),Y(1) is drawn before X(2),Y(2) and so on.
Next, the connections between the points must be considered. This information is provided by a third array, this time a two-dimensional one. The line array is dimensioned W(1 ... 2,1 .. *i*) where *i* is the number of lines to be drawn in the picture. Now the first dimension of the array W indicates that for each line number there are two items of data, as you can see in the next table. These two items of data are not coordinates as such but are what are termed indices (singular: index). An index in computer jargon is merely a pointer to some other piece of information in the computer. In this case each index points to an element number in the X and Y arrays. The first index for each line corresponds to the coordinates of the start point for the line. The second index corresponds to the finish point for the line. So the complete data for drawing a square (Figure 7.7) might look as follows.

i	X(*i*)	Y(*i*)	W(1,*i*)	W(2,*i*)
1	50	150	1	2
2	150	150	2	3
3	150	50	3	4
4	50	50	4	1

Notice that the number of elements in *X*, *Y* and *W* are the same in this example, but this need not always be the case. If the following picture is to be drawn (Figure 7.8), the *W* array would contain a line break as you can see from the data here.

i	X(*i*)	Y(*i*)	W(1,*i*)	W(2,*i*)
1	50	150	1	2
2	150	150	2	3
3	150	125	4	5
4	150	75	5	6
5	150	50	6	1
6	50	150		

Figure 7.7: A Square Showing the Point and Line Data Used in its Construction

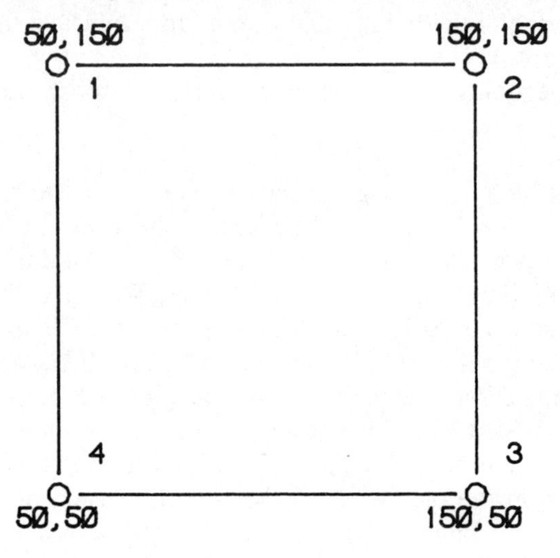

Figure 7.8: A 'Broken' Square Showing Point and Line Data

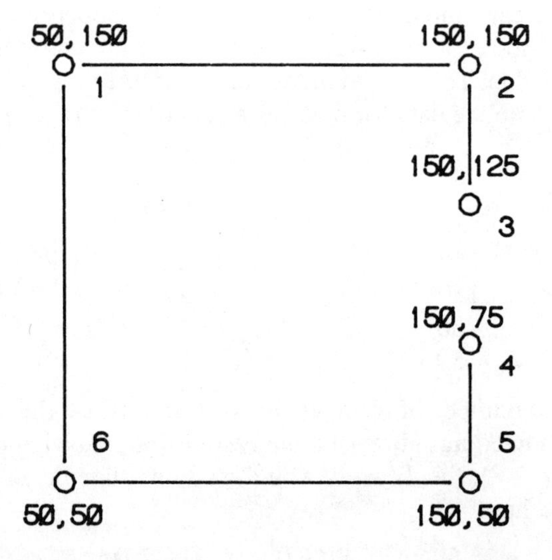

By using the X, Y and W arrays to store the picture information, you can construct any line drawing you wish. If you consider a screen full of information as a single entity constructed of information derived from the X, Y and W arrays, then it is a static set of data which may be very aesthetic to look at, but is restricted in its usefulness. What happens if we want to interact with the picture in some way? Perhaps we wish to move part of the picture to a different location on the screen, or alternatively we may want to erase or amend part of the picture.

In order to consider parts of the picture without the whole, we must introduce a new concept, the picture segment. A picture segment is a section of a picture which may be treated in its own right.

If the total picture is to be treated as one segment, then our X, Y and W arrays are quite sufficient to store and construct the segment. If more than one segment is to be displayed, the input data must be stored in either a series of arrays (X1, Y1, Z1, . . ., Xn, Yn,Zn), or else the segments can be stored as blocks in the single, X, Y and Z arrays. This latter method is neater and more widely used, so this is the one that we will consider. What method do we use to define the segments if they are all in one array? Let us take a pictorial example first.

Figure 7.9 shows the structure of a model of a bacteriophage made up of a number of subunits (the subunit structure is taken from a paper by Thompson and Goel, 1985). We will use this structure as the starting point for our discussion of segment handling, but emphasise that the data representation is not related to Thompson and Goel's own work.

The phage 'segments' are defined as shown in Table 7.1 (note that these simple segments are all made up of rectangles).

Table 7.1: Segment Defining Bacteriophage Subunit

Molecule type	Code letter	Number
Bacterial cell wall barrier molecule	A	many
Molecule for insertion	B	?1
Active molecules	C	1
Active Molecules	D	5
Pressure molecules	E	2
Base molecules	F	2
Base molecules	G	2

The storage arrays for the data making up the segments might appear something like this:

i	X(i)	(Yi)	W(1,i)	W(2,i)
1–4	DATA FOR MOLECULE A			
5–8			B	
9–12			C	
13–16			D	
17–20			E	
21–24			F	
25–28			G	

Figure 7.9: A 'Bacteriophage' Generated Using the Segment Data Structures Described in this Chapter. The general appearance of the phage is taken from a simulation paper by Thompson and Goel (1985). (a) Whole phage showing duplicated segments. (b) Phage with segments removed. Code letters refer to segment types described in the text

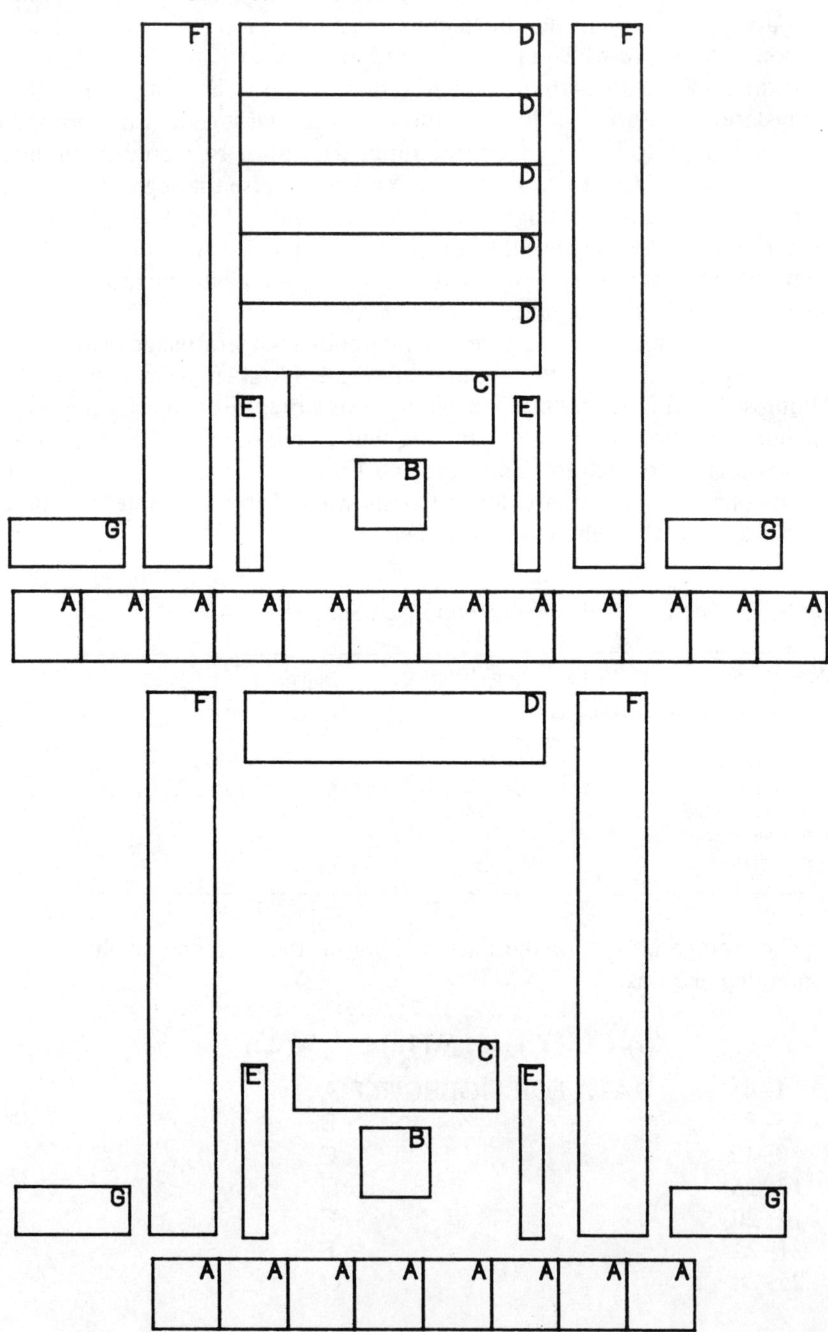

Note that there is only one set of data for each molecule type, but molecules A,D,E,F and G are all present in multiple copies. We therefore need to be able to recall the segment data at will to copy individual segments. In order to access the data representing a particular segment, the start and finish indices for the segment are needed. In other words, we need to know the start point for the first line of each segment, for example the chair, and the finish point for the last line of the segment. To do this we define a new two-dimensional array, which is dimensioned S(2,NS), with NS representing the total number of segments, in this case seven. The appearance of the S array for our example is as follows:

i	S(1,i)	S(2,i)
1	1	4
2	5	8
3	9	12
4	13	16
5	17	20
6	21	24
7	25	28

So S(1,i) is the index of the start line for the *i*th segment and S(2,i) is the index of the finish line for the *i*th segment. You can see that our data structures are getting quite involved! We have a 'three-tiered' system, with the S array pointing to the W array pointing to X and Y arrays (Figure 7.10).

Figure 7.10: Links Between Arrays Used to Hold Point, Line and Segment Data

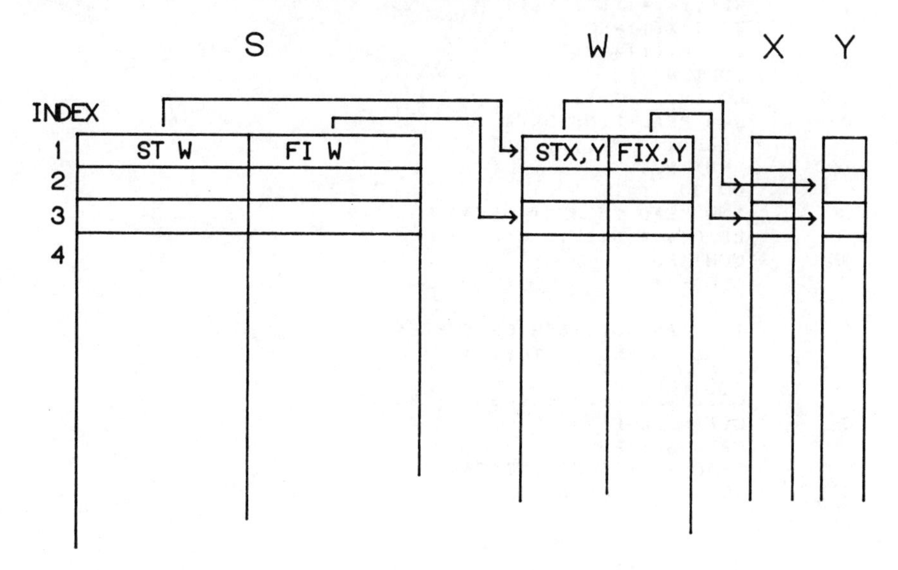

We now have all the information necessary to access the segment from the main program generating the picture. To actually draw a segment at a given position on a display device we need to be able to translate the x and y coordinates for each point in the segment. This transformation step is made especially easy if the original data is calculated around the origin for each segment (see Chapter 4). The Fortran program VIRUS reproduced below (program 7.3) was used to reconstruct the pictures of the phage reproduced in Figure 7.9 above. The point and line data for each segment is held in the data file VIRUS.DAT, while a second data file called VPOSN.DAT holds the number of occurrences of each segment together with the points around which each segment occurrence is to be drawn.

Program 7.3

```
C       THIS IS A FORTRAN DRAWING PROGRAM
C       TO DRAW BACTERIOPHAGE SUBUNITS
C       AS A DEMONSTRATION OF THE USE OF DATA
C       STRUCTURES TO DEFINE POINTS, LINES AND SEGMENTS
C
        DIMENSION X(28),Y(28),W(2,28),S(2,7)
        CHARACTER*1 LABEL(7)
        DATA LABEL/'A','B','C','D','E','F','G'/
C
C       ANY SCALING NECESSARY?
C
        WRITE(5,*) 'INPUT SCALE FACTOR (1.0 FILLS THE SCREEN!)'
        READ(5,*)SC
C       NOW READ COORDINATE DATA FOR EACH SEGMENT
        OPEN(UNIT=4,NAME='VIRUS.DAT',TYPE='OLD')
        DO 100, I=1,28
        READ(4,*)X(I),Y(I)
        X(I)=X(I)*SC
        Y(I)=Y(I)*SC
100     CONTINUE
        DO 200 I=1,28
C       NOW READ LINE DATA
        READ(4,*)W(1,I),W(2,I)
200     CONTINUE
        DO 300 I=1,7
C       NOW READ SEGMENT DATA
        READ(4,*)S(1,I),S(2,I)
300     CONTINUE
        CLOSE(4)
C
C       INITIALIZE GRAPHICS SCREEN
        CALL IGLINI(ID,IO,IB)
C       SET TEXT SIZE
        CALL TXSIZE(0,1.4*SC,1.795*SC)
C       CLEAN PAGE
        CALL NEWPAG
C       READ POSITIONING DATA
C       OC     = NO. OF COPIES FOR GIVEN SEGMENT
```

```
C        XP,YP = COORDINATES FOR CENTRE OF THIS SEGMENT COPY
         OPEN(UNIT=4,NAME='VPOSN.DAT',TYPE='OLD')
         DO 400 I=1,7
         READ(4,*)OC
         DO 500 J=1,OC
         READ(4,*)XP,YP
         XP=XP*SC
         YP=YP*SC
C        NOW PLOT THE SEGMENT COPY
         DO 600 K=S(1,I),S(2,I)
         CALL MOVE(X(W(1,K))+XP,Y(W(1,K))+YP)
         CALL DRAW(X(W(2,K))+XP,Y(W(2,K))+YP)
600      CONTINUE
C        PLOT LABEL FOR THIS SEGMENT COPY
         K=K-3
C        LABEL PLOTTED CLOSE TO UPPER RIGHT CORNER OF RECTANGLE
C        SET DISPLACEMENT FOR TEXT
         DIS=(3.*SC)
         CALL MOVE(X(W(1,K))+XP-DIS,Y(W(1,K))+YP-DIS)
         CALL TEXT(1,LABEL(I))
500      CONTINUE
400      CONTINUE
         CLOSE(4)
         CALL GRSTOP
         END
```

7.5 A DATA STRUCTURE FOR HIDDEN LINES TREATMENT

A simple hidden lines algorithm for convex objects was discussed in Chapter 6. What form must the data be in to use this algorithm? So far we have dealt with arrays X, Y and Z which hold sequential lists of the x, y and z coordinates for the points in the picture. We also have an array W which contains the data for drawing lines between the sets of coordinates. We now need to introduce two new arrays which will be used to hold the surface information. These arrays are called FA and NL. FA is dimensioned FA(i,j) where i is the number of lines defining surface j. The values in each element of the array are the indices held in our W array. Knowing the value of a given element of FA, then, allows us to access the coordinates of the points at either end of the line to which it refers. The second array, dimensioned NL(k), holds the number of lines defining each surface, with k representing the total number of surfaces.

If this is unclear, refer to Figure 7.11 which shows the relationship between all the arrays we have introduced so far: the data is for a cube. It is important that you take the trouble to work out this relationship, or else you will have difficulty in creating your own data sets for hidden lines treatment.

The data for the arrays defining the cube shown in Figures 7.12 and 7.13 is as follows

Figure 7.11: Modification of the Point and Line Data Structures in Figure 7.9 to Hold Surface Data

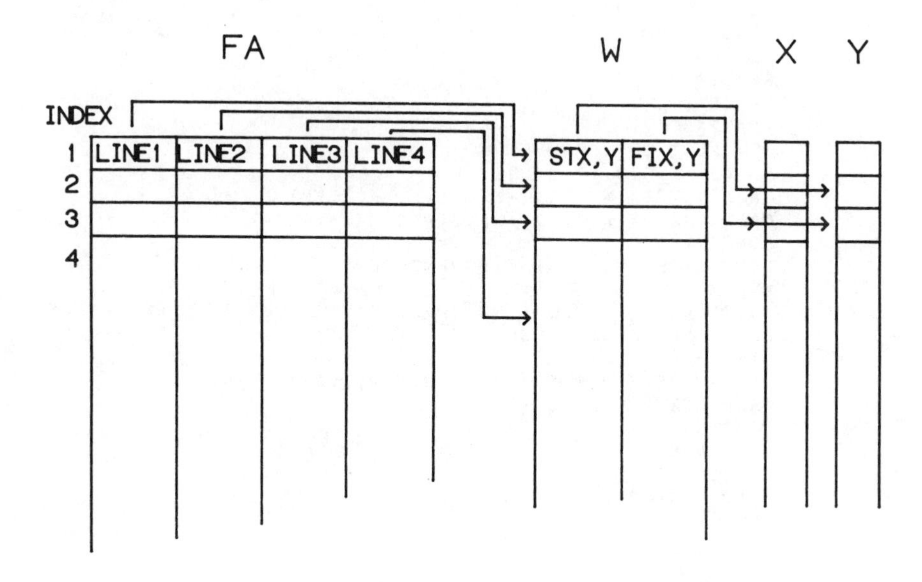

Figure 7.12: A Cube Defined in Three-dimensional Space. Note that the cube is drawn with its centroid at the origin 0,0,0

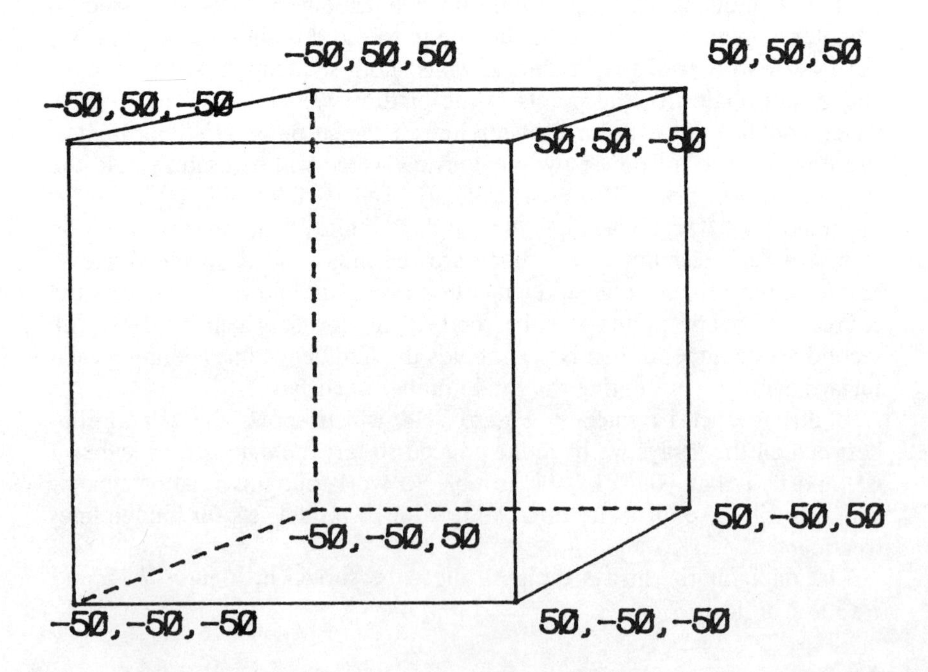

Figure 7.13: Definition of the Point, Line and Surface Data of a Cube. Underlined numerals are lines (that is, indices of the W array). Other numerals are *x*,*y* coordinate points (that is, X,Y array indices). Letters *A* to *F* refer to facets 1 to 6

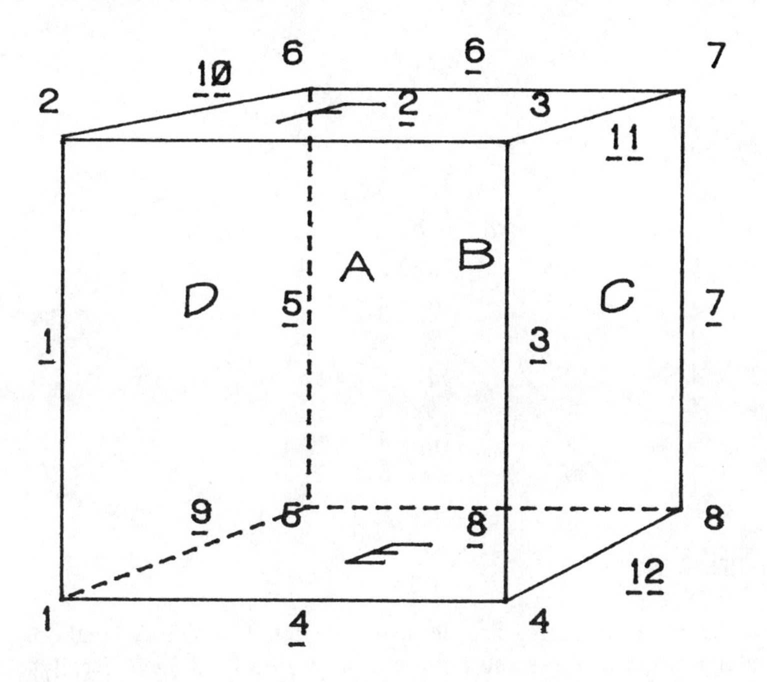

npts = 8

no.	X	Y	Z
1	−50	−50	−50
2	−50	50	−50
3	50	50	−50
4	50	−50	−50
5	−50	−50	50
6	−50	50	50
7	50	50	50
8	50	−50	50

lines = 12

	W	
no.	1	2
1	1	2
2	2	3
3	3	4
4	4	1

5	5	6
6	6	7
7	7	8
8	8	5
9	2	6
10	1	5
11	3	7
12	4	8

no.	NL
1	4
2	4
3	4
4	4
5	4
6	4

	FA			
no.	1	2	3	4
1	1	2	3	4
2	5	6	7	8
3	3	11	7	12
4	1	9	5	10
5	4	10	8	12
6	2	9	6	11

7.6 REFERENCES

Graphics data structures are discussed in detail in Giloi's textbook, and simpler aspects of the same topic are illustrated by Angell. Really simple graph techniques (and full BASIC programs) for implementation on microcomputers are given in Hearn and Baker.

Angell, I.O. *A Practical Introduction to Computer Graphics.* (Macmillan, London, 1981)

Giloi, W.G. *Interactive Computer Graphics* (Prentice Hall, Englewood Cliffs, New Jersey, 1978)

Gribble, F.C. 'Night jars in Britain and Ireland in 1981', *Bird Study* 30 (3) (1983), 164–76

Hearn, D. and Baker, M.P. *Microcomputer Graphics: Techniques and Applications* (Prentice Hall, Engelwood Cliffs, New Jersey, 1984)

Thompson, R.L. and Goel, N.S. 'A Simulation of the bacteriophage assembly and operation' *Biosystems*, 18 (2) (1983), 23–46

8 Reconstruction Methods for Cell Systems

8.1 TISSUE RECONSTRUCTION

It is often difficult to analyse biological material at the level of the whole, be it whole organism, whole organ or whole tissue. If subcellular or cellular patterns are to be studied, the relationships of organelles or cells must be considered. Various methods exist for separating components: at the ultrastructural level centrifugation may be used, and at a higher level microdissection is often possible. There are several problems inherent in the use of these methods.

(1) There is often a loss of structure when removing individual elements from a system: centrifugation causes rupture of cells, and the cell organelles may themselves be disrupted.

(2) Even when the elements are separated, it is not always easy to see how they fit together in the intact system.

(3) Separation methods may not be feasible at all, because of the complexity of the whole system: an example would be neurones in the brain.

A solution to these problems is to make reconstructions from serial sections, a technique that has been a tool in the histologist's armoury since the last century. Reconstructions have been used at all levels of structural analysis, from molecules and cell organelles up to the macroscopic relationships between organs or plant tissues.

'Classical' reconstruction methods use histological sectioning as the data source. There is also an exception that will be considered in the next chapter. This is tomography, where radiation or ultrasound is used to obtain coordinate data.

8.2 THE ROLE OF COMPUTER GRAPHICS

Reconstructions from histological material were for many years fabricated from physical materials: plasticine, cardboard, papier mâché, perspex or even wood. Patience and dexterity are required to use these materials, and even an accurate model of this kind has its drawbacks, the main disadvantages being the solid and permanent nature of the model precluding visualisation of internal structures, together with the difficulty of obtaining quantitative data from the model. Computer graphics can offer a ready solution. By building up the data into a computer image, geometrical transformations (see Chapters 4 and 5) can be performed, and a variety of different treatments like wireframe drawing or solid modelling can be used. Data can be organised into graphic segments so that elements can be displayed, highlighted or temporarily removed during image construction. Data entry into the computer can also be extremely rapid with the use of image analysis techniques. Over the past 10 years computer graphics have been used in a number of different reconstruction systems, and some of these will be reviewed here.

In the present chapter we will look at a range of techniques used in tissue analysis and reconstructions. We will begin by considering methods of data entry, followed by discussion of reconstruction of two and three-dimensional images from histological sections.

8.3 INPUT OF DATA

Practical study of biological systems can never be completely replaced by computer methods, and the first step in any histological reconstruction system is to obtain the raw data. The data in the majority of cases consists of single or serial sections, the latter being obligatory if three-dimensional reconstructions are to be made. If we assume that the data to be sent to the computer consist of outlines of cells or organelles we are faced with the problem of obtaining discrete data points in the form of (x,y) coordinates from what may be a set of continuous outlines in the prepared sections.

What methods can be used to input data from a given section or series of sections into the computer? There are two main possibilities. The first of these is use of a digitizer (these devices were discussed in Chapter 2). The advantage of using a digitizer is low cost. Although it is possible to pay up to £10 000 for a large digitizer, a 12″ × 12″ (30 × 30 cm) device will only cost a few hundred pounds, and will be of sufficient size for most biological applications. The drawbacks to the use of a digitizer are the length of time necessary to input data, together with the difficulty of obtaining more than a crude set of data from sectioned material. A digitizer might be used to input the coordinates of cell outlines by tracing round the outlines with a hand cursor for example, or the relative positions of different classes of cell on a section could be marked.

The alternative method of data entry is the use of a camera as an image capture device (see Chapter 9). A digitized version of the image of the complete section can then be entered into the computer, and equipment for processing either colour or monochrome images is available. The ‘raw’ image consists of a matrix of grey shades or colours which will then need to be processed within the computer. The correspondence between the displayed image and the section allows far more complex data to be obtained than could be inputted using a digitizer. Things are not quite as simple and straightforward as they seem, however. Although the image obtained may allow the experimenter to perform fancy operations like highlighting of areas with particular colour densities, or calculation of areas occupied by particular substructures, image analysis for three-dimensional reconstruction often requires input of outlines only, and the mass of data obtained by photographic input may seem a hindrance rather than a help. Solutions to this problem may be to trace the required outlines and to input the tracings via the image capture device, or possibly to edit the complete image on screen so that only the outlines remain. This method works only if suitable software is at hand.

8.4 TWO-DIMENSIONAL ANALYSES

We have seen that data can be inputted from a digitizer or image analyser, and that the complexity of the data will usually be far less if a digitizer is used. The first question to ask before deciding on a particular analysis system therefore concerns the information to be obtained from the data. If the data is to provide only measurements of distances and areas for statistical analysis, then a relatively simple digitizer-based input system may suffice. By tracing around cell or organelle outlines (for example, like those shown in Figure 8.1) with the digitizer cursor, coded images of the structures can be stored and processed by a computer. The resolution of the stored image will be determined by the number of coordinate pairs inputted from the digitizer, although a smoothing routine may be used to 'tidy up' the image.

Once the data has reached the computer, it has to be processed before most operations can be performed on it. Once a computer representation of the data is available, computer programs can be used to calculate the various measurements from the data. Areas can be calculated from the space occupied by given parts of the image. Calculation of maximum and minimum diameters can also be performed. If particular elements are to be quantitatively studied on different sections, they may be coded, so that statistical operations may be performed, for example, only on elements of a given type. An example might be the calculation of average areas for given elements, or tests of correlation between pairs of element types.

It is quite possible to build a 'home-grown' section analysis system of this type for a relatively low cost, but of course the actual cost (and the time it will take to program the system) will increase as more sophistication is built in. Let us recap on the components required.

Figure 8.1: Input of Cell Images from Photographic Data. The cell outlines highlighted here may be sent to a computer data file via a digitizer. The digitizer cursor is moved around the cell outlines and boundary points are stored

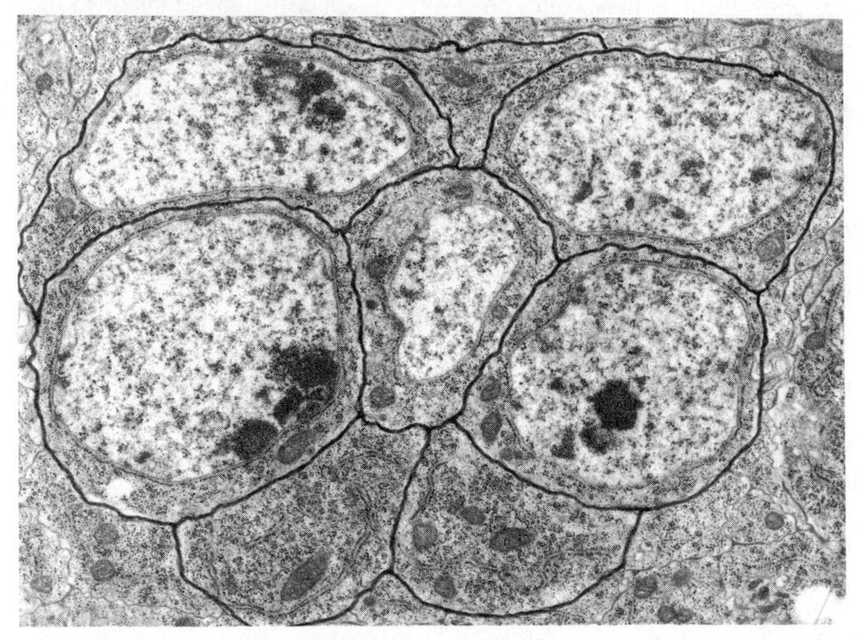

(1) Digitizer.
(2) Computer (with sufficient disk storage to hold data files).
(3) Software to carry out data manipulations and statistical analyses.

You can of course purchase a complete system designed for laboratory use, for a considerably greater financial outlay. As with all computer decisions, factors such as convenience and whether or not a commercial system is tailored closely enough for your needs must be taken into account.

Most of this chapter will be concerned with three-dimensional reconstructions, but analysis of two-dimensional data can itself provide valuable evidence in hypothesis testing. Jordan *et al.* (1985) present an example of such a two-dimensional analysis. The movement of lysosomes within mouse macrophage cells was studied to correlate this movement with the ability of the lysosomes to contact and fuse with phagosomes containing an ingested foreign object. Does lack of movement prevent the lysosomes getting to the phagosomes to take part in the fusion process? The trajectories of individual moving lysosomes are impossible to follow, so a computer graphics technique was used to observe the movement.

The introduction of certain chemicals (such as ammonium chloride, poly-D-glutamic acid) slow or stop motion within normal cells and also inhibit lysosome–phagosome fusion. Cells treated with these chemicals, together with untreated cells were photographed at intervals using phase contrast microscopy, and enlarged prints of the photomicrographs were marked with the positions of the cell boundaries and the lysosomes. Fixed reference points (for example, fat globules) were also marked on each micrograph. The marked elements were digitised and entered into a PDP 11/34 computer. Pairwise comparisons were performed on the data, with the cell outlines aligned as closely as possible using the fixed markers. Colour graphic display (Plate 1) was used to allow checking of lysosomal movement by the following method. Lysosomes present at time $t1$ were displayed as green markers, and those present at time $t2$ as red markers. Superimposition of green and red markers produced a yellow coloration, and the predominance of yellow markers indicated minimal lysosomal movement. The graphics analysis was coupled with a quantitative 'nearest neighbour' analysis to show that the treated cells did indeed show less lysosome movement than the untreated cells.

8.5 THREE-DIMENSIONAL RECONSTRUCTION

Although much information about the structure of biological material can be gleaned from study of single histological sections, a true picture of three-dimensional structure is best obtained by using three-dimensional reconstruction methods. As we saw in the introduction to this chapter, classical reconstruction methods are very slow, tedious and limited in their usefulness. The solution is offered by computer reconstruction methods, but it is not possible at the time of writing to obtain commercial packages to do the tasks involved. Various reconstruction methods have been published, however, and indeed the authors have developed two packages themselves which may be used for reconstruction. These packages will be described below, but we will first consider the requirements for reconstructing images in three dimensions.

The most basic form of these requirements is as follows

(1) Data input device and program (as for analysis of single sections).
(2) Program to transform input data into a form compatible with the reconstruction program.
(3) Reconstruction program.

We need not look at data input, as essentially the same methods may be used as for two-dimensional analysis. The need for a transformation program will depend on the type of reconstruction to be performed. In the simplest case, the outlines of cells or organelles obtained from a digitizer may be directly used in a reconstruction program: this method has been used by Moens and Moens (1981), and is shown in Figures 8.2 and 8.3. In neural analyses (see for example Mazziotta and Hamilton, 1977; Capowski, 1977; CELL program — see below), paths of nerves may be followed through a sequence of sections, and in this case, each neurone is represented by a point in each section, and lines in the reconstruction represent paths of neurones. The digitized data can also be directly used as input to the reconstruction program in this case,

If surfaces are to be constructed between sections, points must be chosen

Figure 8.2: Tracings and Plots of Contours. (a) One section of a series through a metaphase chromosome. (b) Digitized image from (a) with 18 points digitized. (c) digitized image from (a) with eleven points digitized: note the loss of definition with this lower resolution. (d) 'smoothed' version of (b); (e) 'smoothed' version of (c). Reproduced from Moens and Moens (1981) with permission

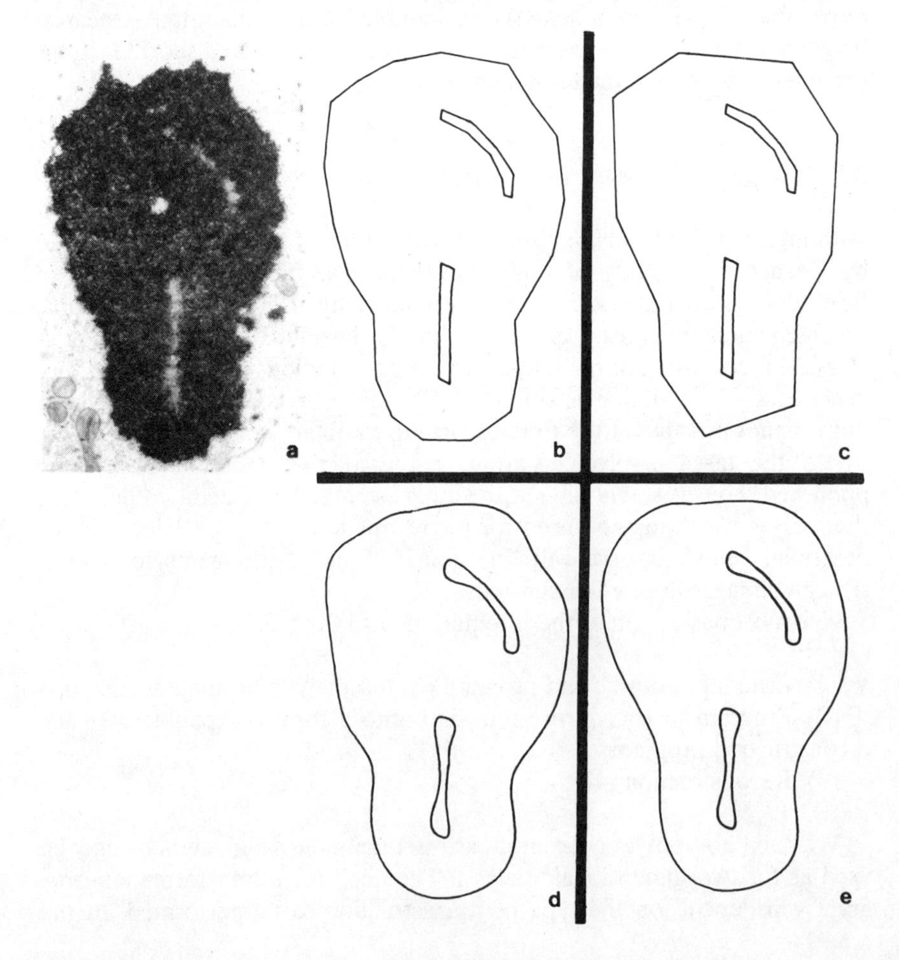

Figure 8.3: Two Stereo Pairs of Contours from Serial Sections through the Metaphase Chromosome shown in Figure 8.5. (c) and (d) Represent a 45 degree rotation of the data in (a) and (b). Reproduced from Moens and Moens (1981) with permission

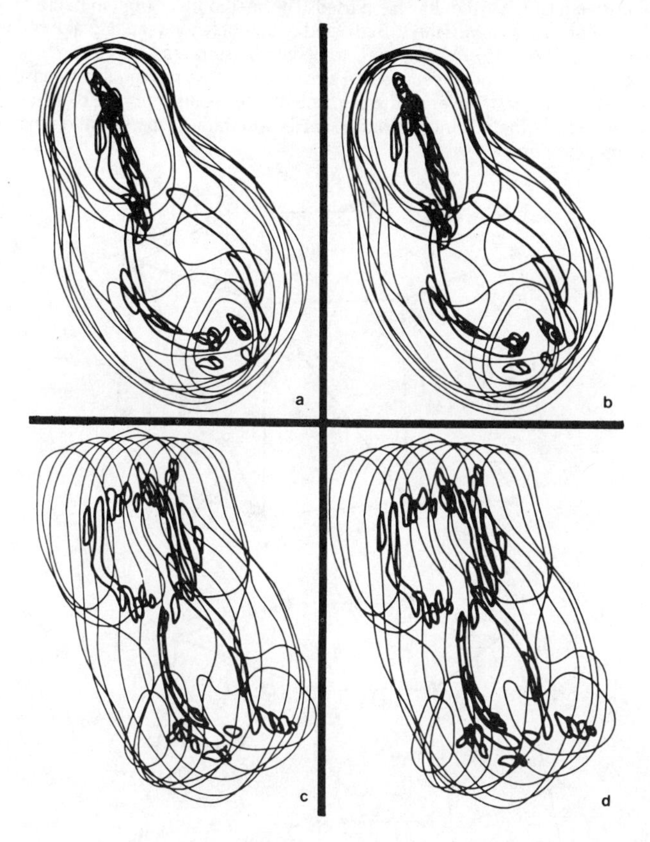

on adjacent sections to identify the polygonal surfaces. In the simplest possible case, surfaces for reconstruction of a single object could be identified by use of the same number of points to define each section, with the line joining each sequential pair of points representing the upper or lower boundary of the surface (Figure 8.4). This arrangement is inflexible, because extra points may need to be digitized if a section is particularly complex, the alternative being to digitize a constant but large number of points on each section.

An alternative method that does require modification of the data is the 'keychain' method devised by Shantz and McCann (1978). A computer algorithm tests the curvature around the shapes generated from the digitized data and only saves the boundary points where the curvature exceeds a predetermined value (Figure 8.5). This method frees the experimenter from worrying about maintaining a constant number of points per section, if individual points are inputted from a digitizer. Shantz and

Figure 8.4: Two Ways of Reconstructing Surface Information from Serial Sections. The two sections *s*1 and *s*2 may be viewed as stacked sheets (see for example Plate 3, but a more realistic image can be reconstructed by reconstructing the surface between the sections. In the top diagram, a triangulated system has been used. This method was used in neurone reconstructions by Shantz and McCann (1978), and allows polygon mapping algorithms (see Fuchs, Kedem and Uselton, 1977) to be used to define the surfaces. With this method, an arbitrary number of points may be digitized on each section. The rectangular reconstruction method (bottom) is more restrictive, as the same number of points must be digitized on each section. This method has the advantage that a simple 'wireframe' representation may be built up for little programming effort

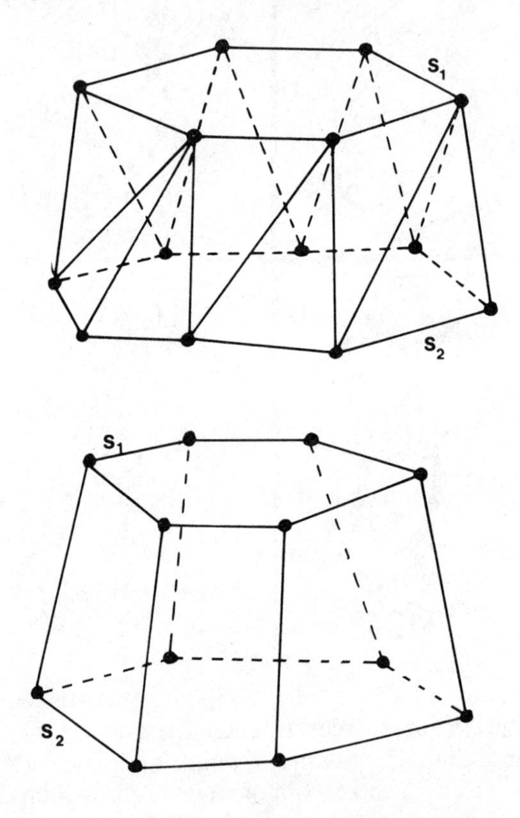

McCann used a digitizer with a tracing pen to input data, and a continuous stream of coordinate data was inputted as the pen traced round each outline.

Some reconstruction programs, for example RECON described below, essentially display a stack of sections in a similar 'bacon slice' manner to the Moens and Moens program mentioned earlier. The difference between RECON and the Moens' work is the use of hidden surface removal to aid the analysis of the displayed image (note that the Shantz and McCann system also uses hidden surface removal, but on the surface polygons interpolated between adjacent sections). How must the data be modified for use

Figure 8.5: Use of the 'Keychain' Method for Reducing the Number of Data Points to be Held by the Computer for Three-dimensional Reconstructions. The top diagram represents a digitized cell outline with each digitized point at the same distance apart. A large amount of data is therefore needed to represent the section contour. The lower diagram represents the data after application of a 'keychain' algorithm to reduce the number of data points on boundary regions with low curvature. The keychain concept is discussed further in Shantz and McCann (1978)

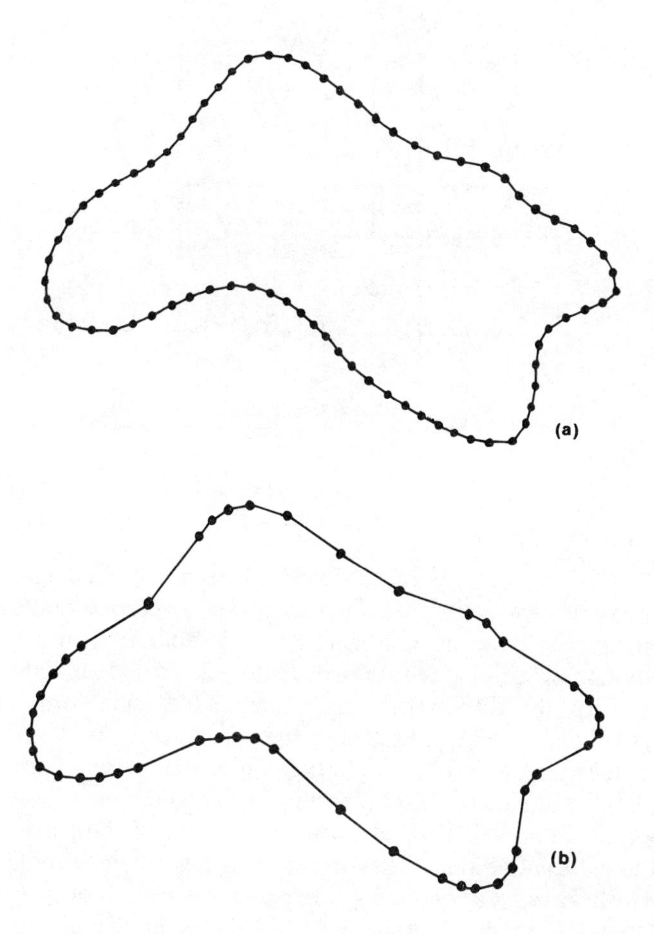

in the 'bacon slice' type of hidden surface program? Recall that the digitized data is in the form of a set of points that are joined by lines to define a surface. The computer 'sees' the objects as sets of lines, and not as surfaces. Data modification takes the form of defining the surfaces present on each section, and in practice this means 'blocking in' outlines. In RECON, we use an integer 80 × 80 array to hold the 'blocked in' image from each single section (Figure 8.6). The three-dimensional reconstruction is therefore performed on two-dimensional planar objects and not on wireframe outlines.

Figure 8.6: Relationship of Digitised Coordinate Data Representing a Cell Outline in Section Together with an Integer Array Used to Hold the Blocked-in Silhouette of the Cell Section

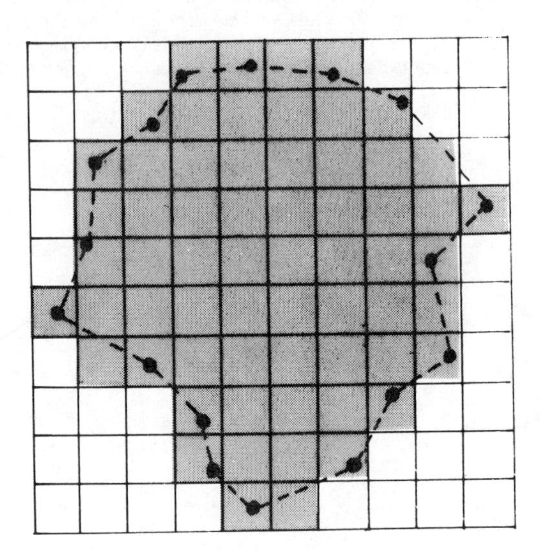

Reconstruction Programs

Once the data from a set of sections has been modified into the form required for reconstruction, the final stage can be carried out. The reconstruction may be done in monochrome or in colour; with a 'wireframe' representation or with hidden surfaces removed. To add an additional level of complexity, the displayed surfaces may have some form of shading applied to them. The data itself and the questions to be asked of it will largely determine the level of sophistication needed in any particular case, but availability of suitable display equipment and computer power will also be important. You will have already got an idea of the sorts of systems needed to cope with hidden lines and shading algorithms from Chapter 6.

McIntosh *et al.* (1979) used three-dimensional reconstruction techniques to visualise the central mitotic spindles in the diatom *Diatoma vulgare* (Figure 8.7). The aim of this work was to provide data for a model describing microtubule dynamics during mitosis. Thin sections were cut through Epon-embedded cells, and the digitizer described by Veen and Peachey (1977) was used to input two-dimensional coordinates of the polar microtubules of which the mitotic spindles are made. Three-dimensional data were built up by successive digitization of a number of adjacent sections, and the data was stored on disk using a CDC 6400 computer. Display was performed on a CDC 774 interactive graphics terminal, and the following sequence of steps was used to analyse the material.

Figure 8.7: Stereo Pairs of Prometaphase (Top), Early Metaphase (Centre) and Metaphase (Bottom) Central Spindles of *Diatoma vulgare*. Reproduced from McIntosh *et al.* (1979) with permission

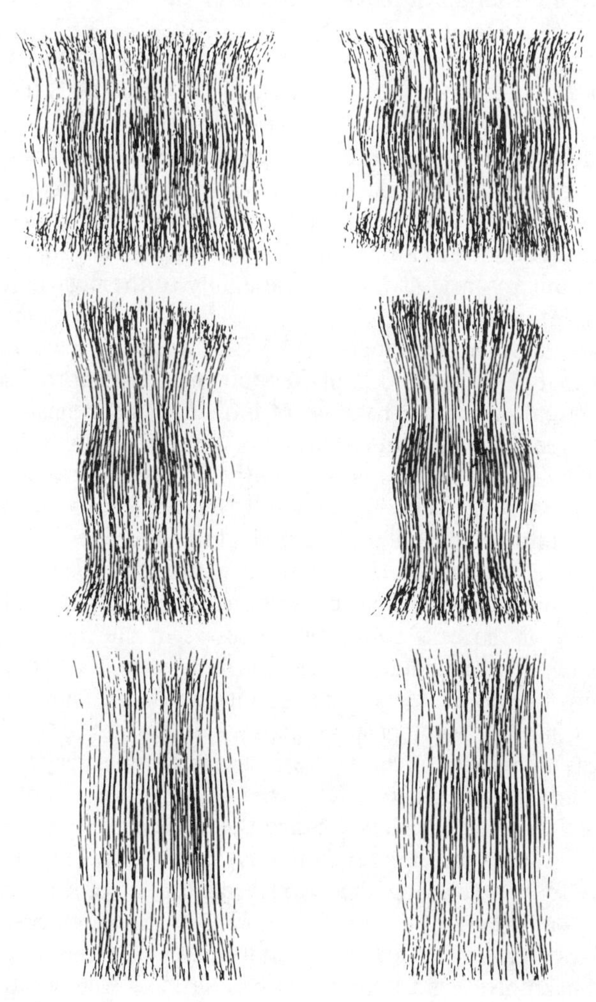

(1) Section alignment.

(2) Data checking.

(3) Calculation of numeric data (such as microtubule density, microtubule number versus position along the spindle).

(4) Display of three-dimensional structure as a two-dimensional projection: the least squares method was used to smooth sharp bends caused by digitizing errors, electron optical aberrations and variation in section projection caused by microscope stage tilting.

(5) Output on film or a Calcomp plotter.

Typical spindle reconstructions are shown in Figure 8.7. The use of stereo projection allows a true three-dimensional image to be seen and is a valuable aid to the analysis of this kind of material.

8.6 THREE-DIMENSIONAL RECONSTRUCTION OF NEURONES (CELL)

The way in which neurones are interconnected within the nervous system is one important correlate of animal behaviour. Since this interconnectivity is achieved through the spatial distribution of neural processes (axons and dendrites), the definition of a neurone's three-dimensional shape provides important preliminary insights into the capability of the nervous system to process neural information.

A group of computer programs called CELL has been developed in the author's laboratory by RJM and Paul Gabbott (now at Oxford University), enabling the visualisation and rotation of individual neurones and groups of neurones to identify their three-dimensional morphology.

The first step in the process is to identify neurones by one of several histological methods — Golgi impregnation (Fairen, Peters and Saldanha, 1977), or the intracellular filling of individual neurones with markers (Martin, 1984). Tissue containing identified neurones is then sectioned at an appropriate thickness (20–200 μm) and is subsequently stained to reveal the morphology of the cells and their processes in the light microscope. With the aid of a camera lucida, the distribution of the axons, axonal swellings (known to be the loci of specific intraneuronal connections), and dendrites seen in each histological section are drawn in the x,y plane onto separate sheets of paper. Each sheet of paper therefore represents each section by a 'histological plane'. Hence the cell is representated by overlapping parallel histological planes. Since the system of processes is continuous, 'fiducial markers' (reference points) can be carefully drawn onto each sheet of paper enabling the correct alignment of the histological planes. Using a program called CELL.PLOT in conjunction with a Summagraphics digitizing tablet, the xy coordinates of axons and dendrites appearing on each histological plane are then defined — in addition the xy positions of at least three non-congruent fiducial markers are then recorded. CELL.PLOT also allows the user to label specific parts of the arbors (for example, dendrites versus axons, processes from different neurones, parts of processes in given regions of the brain, or important groups of axonal swellings) so that they can be rotated separately or identified during the rotation of the whole cell. The number of coordinate points used to digitize a process depends on the resolution desired.

Three programs (CELL.RECON, CELL.UNFOLD, CELL.ROTATE) are used to allow reconstruction and rotation of the neurone(s) in the following ways.

Planar Rotation

In this reconstruction and rotation procedure, each histological section is assumed to be an infinitely thin plane, with adjacent planes separated by a user-defined value (t) — usually the section thickness. This value defines the z coordinate (0, t, $2t$, $3t$) of each histological plane. (If necessary, the scaling of the x, y and z coordinates may be corrected for shrinkage effects during the fixation of the tissue.) CELL.RECON then aligns the histological planes by superimposing the reference points, and also translates the XY coordinates of cellular processes into the same coordinate area. The histological planes within the area are then separated equally by the specified z coordinate. The whole frame of histological planes can then be rotated using CELL.ROTATE through any desired range of x,y,z polar angles to visualise the three-dimensional morphology of the neurone(s) as shown in Figure 8.8. Additionally, parts of the neurone defined by the user may be rotated separately (Figure 8.9). Hardcopies of the rotated image can be obtained using a colour graphics plotter, with labelled portions of the neurone identified by different colours.

Figure 8.8 Planar Rotation using the CELL Program. (a) Distribution of the processes of a layer V pyramidal neurone in the cat visual cortex (area 17) showing the dendritic arbor and the distribution of individual axonal swellings on its axonal arbor. Rotation — *X,Y,Z*: 0,0,0. (b) Rotation *X,Y,Z*: 35,0,90. Here the dendritic arbor shown by the thick lines appears as a series of parallel planes, as does the distribution of the axonal swellings. (c) Rotation *X,Y,Z*: 35,90,0. Scale bar: 100 μm

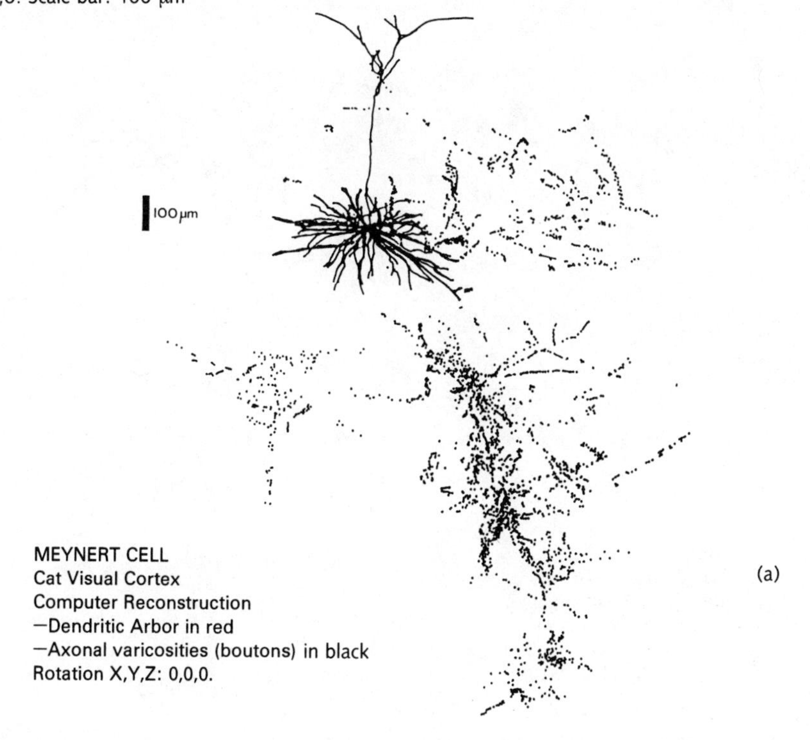

Figure 8.8 continued

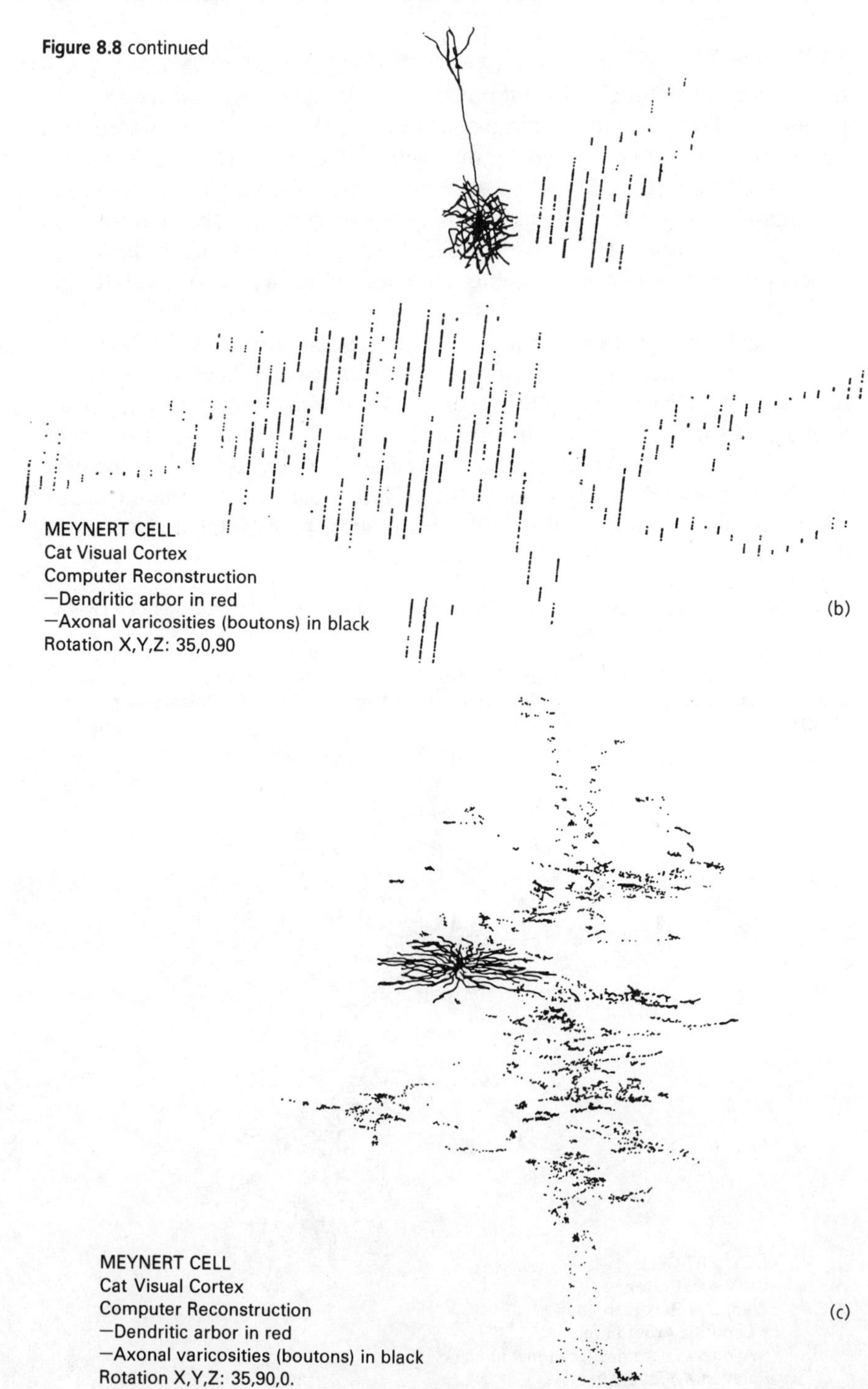

MEYNERT CELL
Cat Visual Cortex
Computer Reconstruction
—Dendritic arbor in red
—Axonal varicosities (boutons) in black
Rotation X,Y,Z: 35,0,90

(b)

MEYNERT CELL
Cat Visual Cortex
Computer Reconstruction
—Dendritic arbor in red
—Axonal varicosities (boutons) in black
Rotation X,Y,Z: 35,90,0.

(c)

Figure 8.9: Interpolated Rotation of the Dendritic Arbor of the Cell in Figure 8.8(a). Rotation about the *Y* axis from 0 to 90 degrees with *X* and *Y* remaining at 0

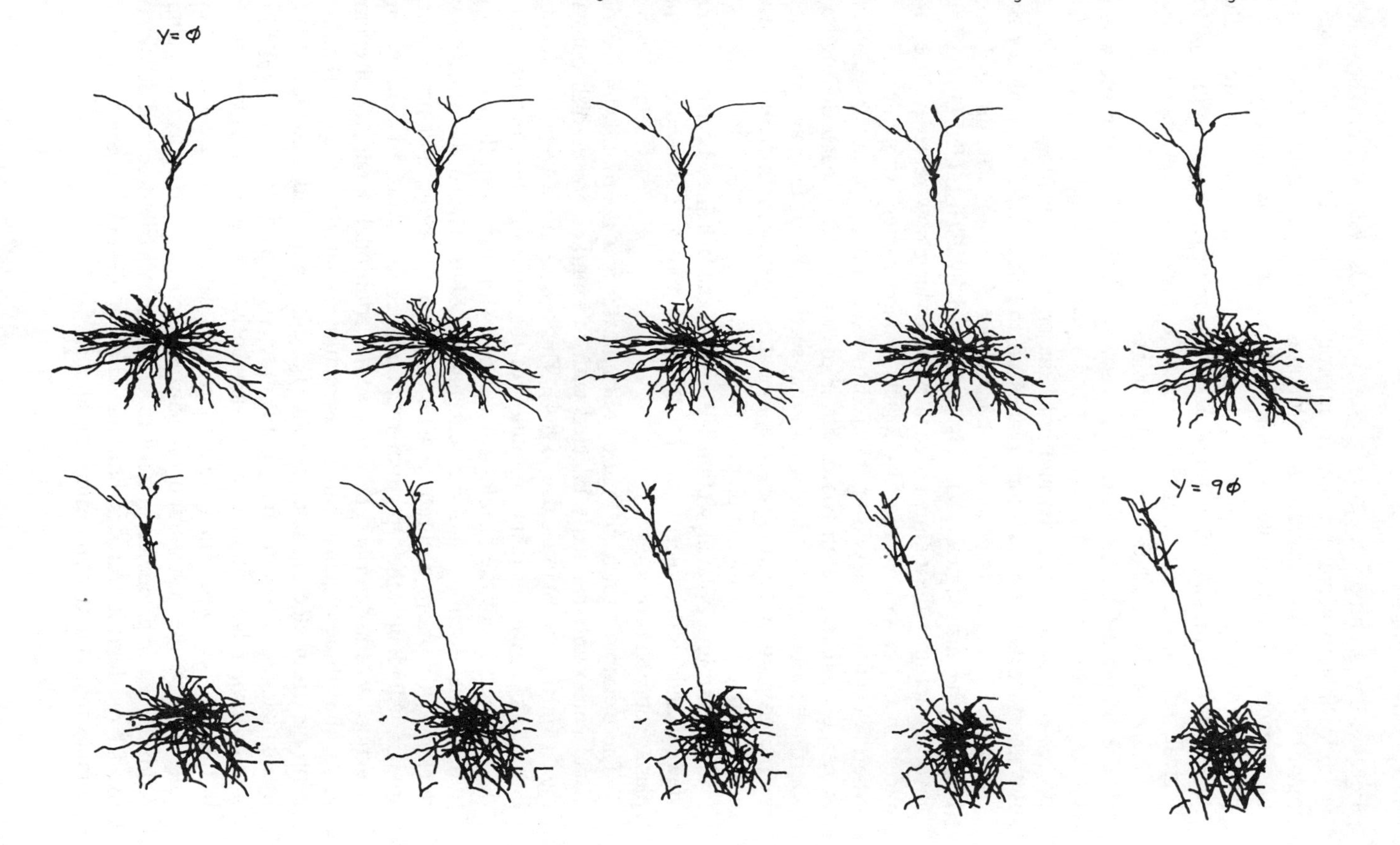

Interpolated Rotation

Because the neural processes exist as a continuous system within the tissue, it is possible to achieve a limited interpolation between the histological planes by ascribing 'pseudo' *z* coordinates to the processes. This can be done because it is possible to identify the points of continuity for a given process from one plane to another. Although only the *xy* coordinates of the processes are digitized, the point at which a process leaves one histological section and enters another can be defined by their similar *xy* coordinates in the translated coordinate frame.

The program CELL.UNFOLD starts with the histological plane where the cell body of the neurone is known to exist (and from where all processes radiate outwards) and links the ends of individual processes at the interface between adjacent planes on the basis of coincident *xy* coordinates using a strict probabilistic matching algorithm. The interpolation between the histological planes is performed by CELL.UNFOLD. This involves defining the points on each plane (which now represent the surfaces on either side of the histological sections, each with thickness (*t*) where a cellular process is considered to have entered or left the section. A perpendicular *z* depth coordinate is then ascribed to each digitized *xy* coordinate along the process within that section. Error flags are generated by the program indicating where ambiguities exist in the matching of processes between histological planes.

Once the interpolated section is finished, the cell may be rotated and viewed, and hard copies of the rotated view may be obtained as for the planar rotation procedure.

The example shown in Figures 8.8 and 8.9 is a neurone from the cat visual cortex intracellularly injected with the marker horseradish peroxidase (Martin and Whitteridge, 1984). The cell ramified over 60 sections 80 μm thick. A limit to the resolution of the reconstruction procedures is the thickness (*t*) of the histological sections with respect to the overall size of the spatial distribution of the neuronal processes. The smaller the section thickness, the more accurate will be the reconstruction, and preliminary studies will define the optimum section thickness required. The reconstruction system described here allows the additional possibility of examining the three-dimensional ultrastructure of parts of the identified neurones from electron micrographs (Fairen, Peters and Saldanha, 1977) by digitizing specific ultrastructural features (such as dendrites, presynaptic and postsynaptic elements) and subsequently rotating the digitized planes in three dimensions. Finally, the coordinate data may be used to statistically analyse the distribution of neuronal processes to detect any similarities between given neural types regarding the distribution and density of axonal or dendritic arbors both in adult or developing animals, or in animals reared under abnormal conditions.

8.7 THREE-DIMENSIONAL RECONSTRUCTION OF NON-NEURAL TISSUE

The analysis of neural tissue is to a large extent concerned with the connections of line data in three-dimensional space. Most cell types have more complex boundaries to consider in two-dimensional section, and so require different reconstruction systems. We will next look at such a reconstruction package developed by one of the authors (RR). This package exemplifies the way in which the questions to be answered about particular biological systems and the availability of particular items of hardware dictate solutions to program development. This system, RECON, is based on a 'baconslice' hidden surface program, and is similar to the TROTS system developed earlier by Veen and Peachey (1977). RECON can display up to 20 different 'objects' from a series of serial sections, and each object can be displayed in a different colour if a suitable colour device is used.

RECON was originally devised to study the spatial interactions of cells during retina formation in the fruitfly *Drosophila*. By making serial reconstructions at various stages of retina development, a series of 'snapshots' may be built up into a progression of cell interactions in this system. The raw data for the reconstructions are a series of transverse thin sections through *Drosophila* eye/antennal imaginal disks, with cell outlines traced for the digitization step (see Figure 8.1). Data input was performed using a Calcomp 9000 series digitizer with a constant number of points digitized per section.

RECON itself is divided into four main sections:

(1) Checking routines: these allow visual checking of the digitized data section by section.

(2) Manipulation routines: these routines allow scale changes and spatial separation of elements of the picture.

(3) Interpolation routines: these allow interpolation of intermediate section levels to smooth the picture.

(4) Display routines: these carry out the hidden surface removed display of the picture, together with geometric transformations and selective display of picture elements.

Checking Routines

The checking routines take data directly from one of the files created from the digitized data. Each file contains a list of the x,y coordinates for each digitized point on each of a series of sections. The coordinate data is headed by the following variable values.

Data item number	Variable
1	Number of sections inputted (NS)

2	Number of cells per section (NC)
3	Number of points defining each cell (NP)

The coordinate data follows in a series of (x,y) pairs, so the sequence is:

Data item number	Variable
4	x pt 1
5	y pt 1
6	x pt 2
7	y pt 2

and so on.

Given the information at the head of the file, it is a simple job for a program to reconstruct the outlines of the cells in each section. Given the variables NS, NC and NP, the first point in section N will be defined by the expression:

$$PT = [(N-1)(NC(NP*2))+4]$$

A Fortran routine to draw this section would therefore look like this

```
C       PART OF ROUTINE SHOW
C       SECTION NO. = NS
C       NO OF PTS/CELL = NP
C       NO OF CELLS/SECTION = NC
C       OPEN SEQUENTIAL DATA FILE TO
C       READ DATA. (CHANNEL = 4)
C       GET TO CORRECT START POINT IN FILE
C
        START = (N-1) * (NC * (NP * 2)) + 4)
        DO 100 I = 1, START
           READ (4,*) VALUE
           IF (I.EQ.3) NC = VALUE
100     CONTINUE
C
        DO 200 I = 1, NC
           READ (4,*) XVAL,YVAL
           CALL MOVE (XVAL,YVAL)
C
        DO 300 J = 1, NP-1
           READ (4,*) XVAL,YVAL
           CALL DRAW (XVAL,YVAL)
```

```
C
300     CONTINUE
200     CONTINUE
```

The SHOW routine will draw the section. In the full RECON package, a grid is also drawn to show the accurate position of the plotted coordinates in the display space (Figure 8.10), but as this merely consists of drawing a series of horizontal and vertical lines it is left as an exercise for the interested reader.

SHOW enables a quick check on the form of the data, enabling the user to see if the section has been digitized satisfactorily, and if the resolution of the image is high enough (that is, are more points needed for each section). If all is well, any manipulations needed can be carried out.

Manipulation Routines

The form that the possible manipulations take will depend on the particular system to be reconstructed. The *Drosophila* imaginal disk reconstructions for which the package was designed are concerned with the analysis of the interactions between the eight photoreceptor cells present in each omma-

Figure 8.10: Digitized Outline of the Electron Microscopy Section Given in Figure 8.1; Eight Cells Have Been Coded

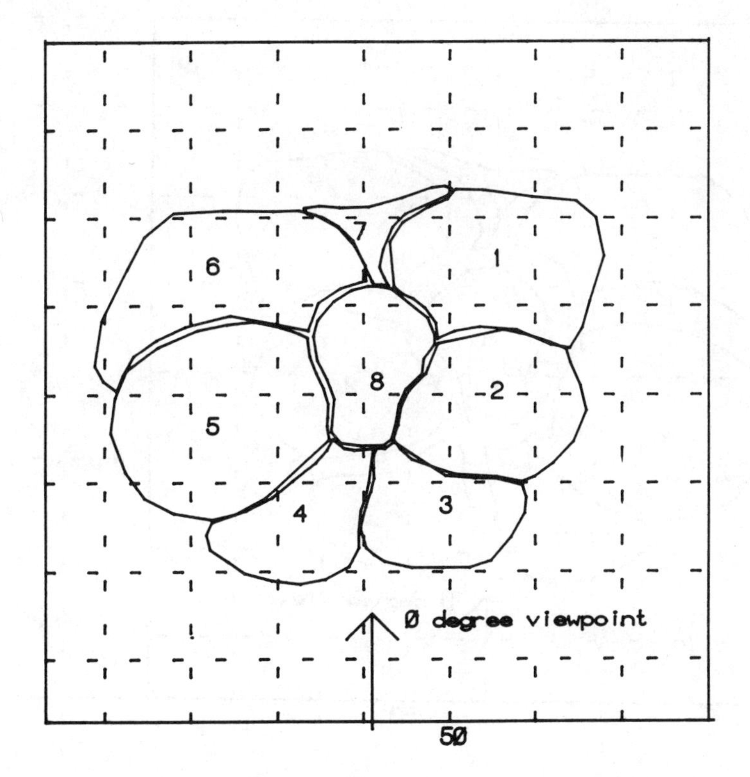

tidium in the developing retina. These cells are closely packed, and a useful option is therefore to observe the cells in an 'exploded' format, with artificial separation between the cells. The coordinate data was manipulated to move the cells apart (Figure 8.11). The separation was achieved by use of a short routine to change each (x,y) coordinate pair by a constant amount. As seven of the eight photoreceptor cells are arranged around a central photoreceptor cell, the coordinates may be adjusted in the following way (note the cell numberings on Figures 8.10 and 8.11):

Cell 1	xnew = xold + 2
Cell 2	xnew = xold + 2
	yold = yold − 1
Cell 3	xnew = xold + 1
	yold = yold − 2

and so on

Figure 8.11: Digitized Outline of the Electron Microscopy Section Given in Figure 8.1; the Individual Cells Have Been 'Separated' to Aid Analysis

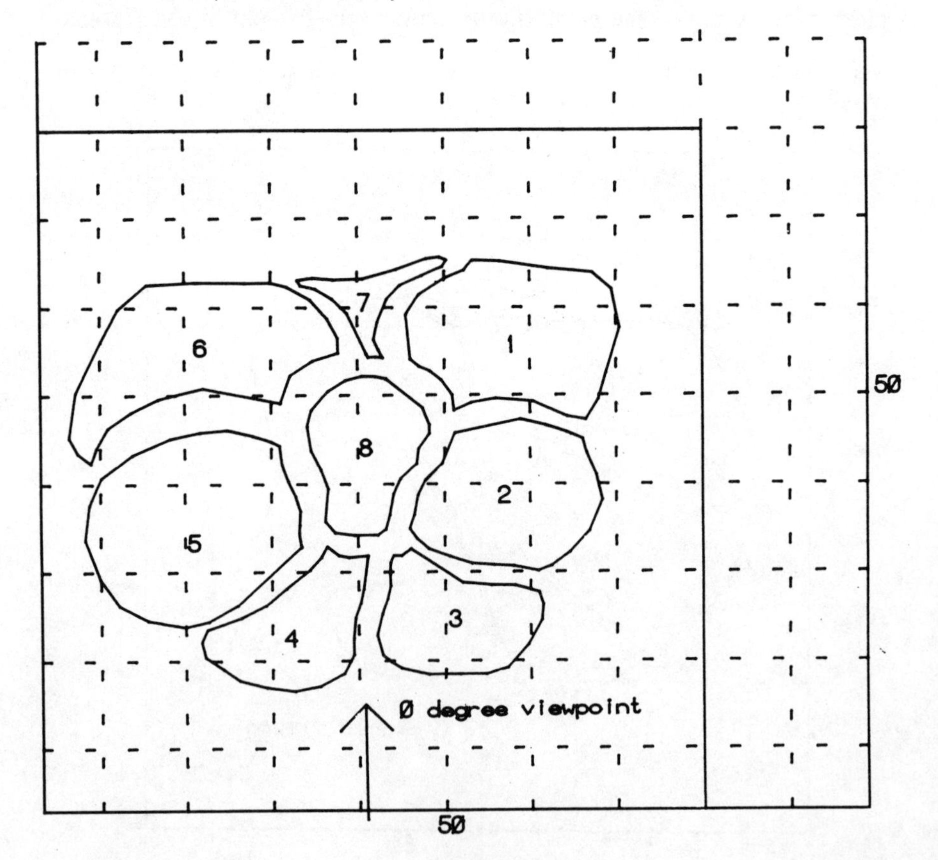

Interpolation Routines

Serial sectioning can be a time-consuming task, and use of the 'baconslice' reconstruction method requires a large number of sections to produce a smooth reconstructed image. A set of routines were therefore added to RECON to carry out interpolation of new sections between each existing pair of adjacent sections. The algorithm can be used recursively to interpolate further sections as required, and the method is simply to average corresponding (x,y) coordinate pairs on the two sandwiching sections, as shown in Figure 8.12.

Display Routines

We now arrive at the display routines. As we saw earlier in this chapter, it is relatively simple to take the raw digitized data and to produce a three-dimensional 'wireframe' representation. This method would produce an unrecognisable mess if we attempted to use it to observe a group of eight cells.

RECON uses the hidden lines algorithm and three-dimensional perspective reconstruction described by Wright (1974). The method used involves four main subroutines. Two routines (INIT3D and SETORG) set

Figure 8.12: Method for Section Interpolation: *x* and *y* Coordinates of Points on Adjacent Sections are Averaged to Obtain the *x,y* Coordinates of Points on the Interpolated Section. In order to limit errors due to poor alignment of points on adjacent sections it is necessary to digitize all points in the same direction (that is, clockwise), and with similarly positioned start points. The interpolation program will also scan the points on the 'lower section' and will use one of these for the averaging process, if one is closer to the point on the 'higher section' than the target point

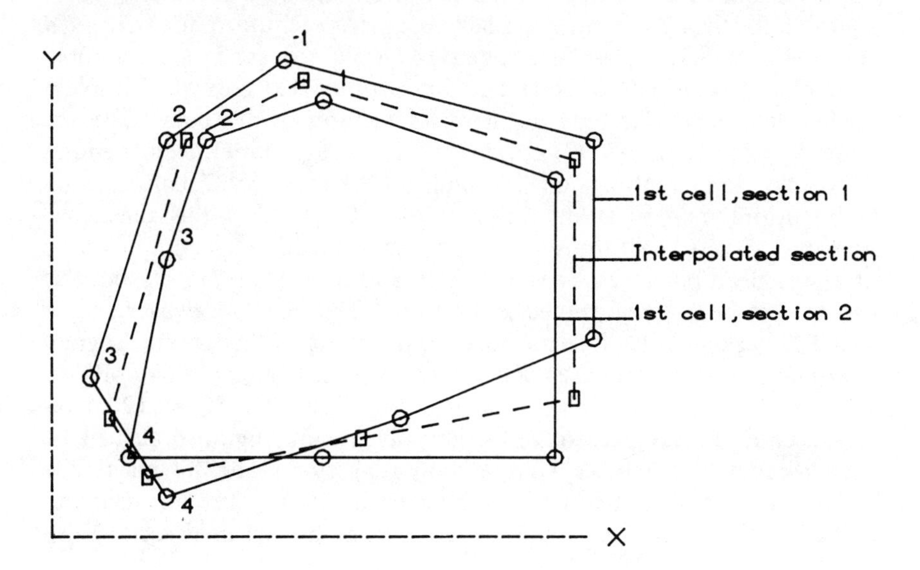

up the line of sight array which defines which three-dimensional coordinate locations are visible from any given line of sight. As this array takes some time to set up and is generated independently of the main program reconstruction data, a program version with INIT3D and SETORG was only run to set up new line of sight arrays for different observer positions. A series of line of sight arrays were stored to be accessed by the 'utility version' of the display program.

The third routine (FILLIN) called by the main program 'blocks in' each section, using the method shown in Figure 8.6. That is, the section is mapped onto an 80 × 80 integer array, and words are set to *n* where a given cell is present and 0 where it is not (*n* is the code for the given cell). The fourth routine (DANDR) draws successive parts of the three-dimensional picture, processing planes further away from the observer in succession.

The main display program also has facilities for removing cells or elements and rotating the picture. Cell removal is effected by excision of the data for the cells to be removed at the stage of input from the main data file. Rotation is performed by matrix transformation around the central (that is, 40,40) point on each 80 × 80 section. As individual cells (or organelles) can be assigned a numeric code, colour display is possible, limited only by the number of colours available on the display device. A simple differentiation is also made possible in monochrome by using different dot patterns.

The suite of programs in RECON has proven to be very time-effective in use. Using the Calcomp 9000 series digitizer, entry of 30 sections each of eight cells with 25 points per cell takes around two hours, and subsequent manipulation prior to viewing can take as little as 15 minutes. Once digitized, the data is of course available in a permanent form for analysis at leisure. Figure 8.13 shows the standard of output produced on a Tektronix 4663 plotter, and Plate 2 shows a colour picture (Tektronix 4027 display) of the same data. The drawing times for pictures of this complexity are three to four minutes (Tektronix 4010 storage tube), ten minutes (Tektronix 4663 plotter) and 25 minutes (Tektronix 4027 colour raster CRT). More modern colour displays should cut down this time considerably. A version of the Wright algorithm claimed to speed up plotting times has been published by van Swieten and de Hosson (1979), but this has not yet been implemented in RECON. The major drawback of the RECON package is its huge memory requirements. Although the program is written in Fortran and uses only simple graphics commands like MOVE and DRAW, each line of sight array occupies 80 × 80 × 80 = 512 kbytes of space. Higher degrees of resolution increase this figure dramatically. Only computer systems capable of supporting storage of this magnitude together with the 120 kbytes needed by the suite of programs together with 640 kbytes for array storage will be usable. The hidden lines section also

Figure 8.13: Different Orientations of the Final Picture of the Cell Group from which the Section in Figure 8.1 was Taken. This is a monochrome reproduction obtained using a Tektronix 4663 plotter, with some cells selectively erased. The central 'axis' position of cell 8 is especially brought out by this reconstruction technique. (a) Cells 2,5 — view orientation = 0 degrees; (b) cells 3,7 — view orientation = 270 degrees; (c) cells 1, 4 and 8 — view orientation = 305 degrees.

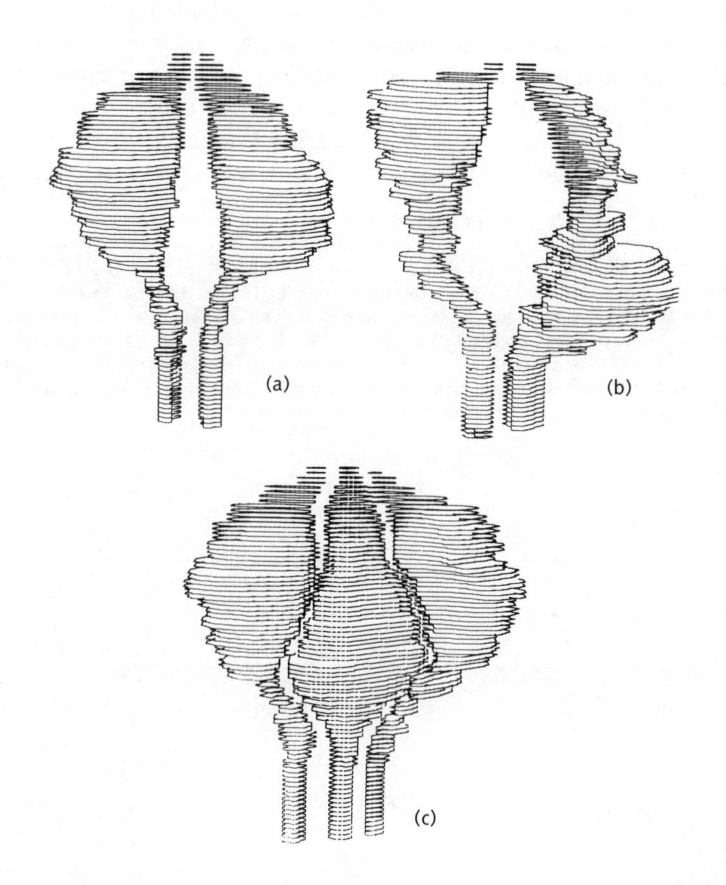

works by bit masking on at least 32 bit words, so the package is not really suitable for implementation on microcomputers: we use a DEC VAX 11/780 to run the package.

RECON has been used extensively in our laboratory to study the development of the third larval instar retina in the fruitfly *Drosophila melanogaster* (Tomlinson, 1985). Each ommatidium in the retina is made up of eight receptor cells plus associated pigment and cornea cells. Serial sections cut at various stages during the third larval instar were therefore used to trace the outlines of receptor cells comprising the precursors of a single ommatidium, these data being used as input to RECON.

Several operational restrictions should be noted. Firstly, the orientation of section cutting is important. Each section is placed adjacent to the next higher section during reconstruction, and alignment is performed by noting landmarks on adjacent sections. Figure 8.14 shows the error that can be introduced if sectioning is not in the required plane.

Section thickness is also important. Ideally, sections should be cut at the required separation for reconstruction. It is, however, possible that variations occur due to lost or badly cut sections (section thickness itself can of course be monitored by section colour). A record must be kept of any variations in distance between adjacent sections. Such variations cannot be allowed for by adjusting the distance between sections in the

Figure 8.14: Errors Introduced by the Reconstruction Method Used in RECON. Sections should be cut transverse to the longitudinal structure to be reconstructed as in (a). Transverse sections are here shown by letters *A-B*, *C-D*, and *E-F*. As landmarks on adjacent sections are used for alignment, and are much easier to align vertically, gross errors will be produced if tangential sectioning is done as in (b). Tangential sections are shown as lines marked by letters *G-H*, *I-J*, *K-L*. In this latter case, alignment of landmarks along the dotted line may lead to the 'worst case' reconstruction shown in (c). Sectioning of (b) along the transverse axis perpendicular to the dashed line in (b) would produce a better reconstruction.

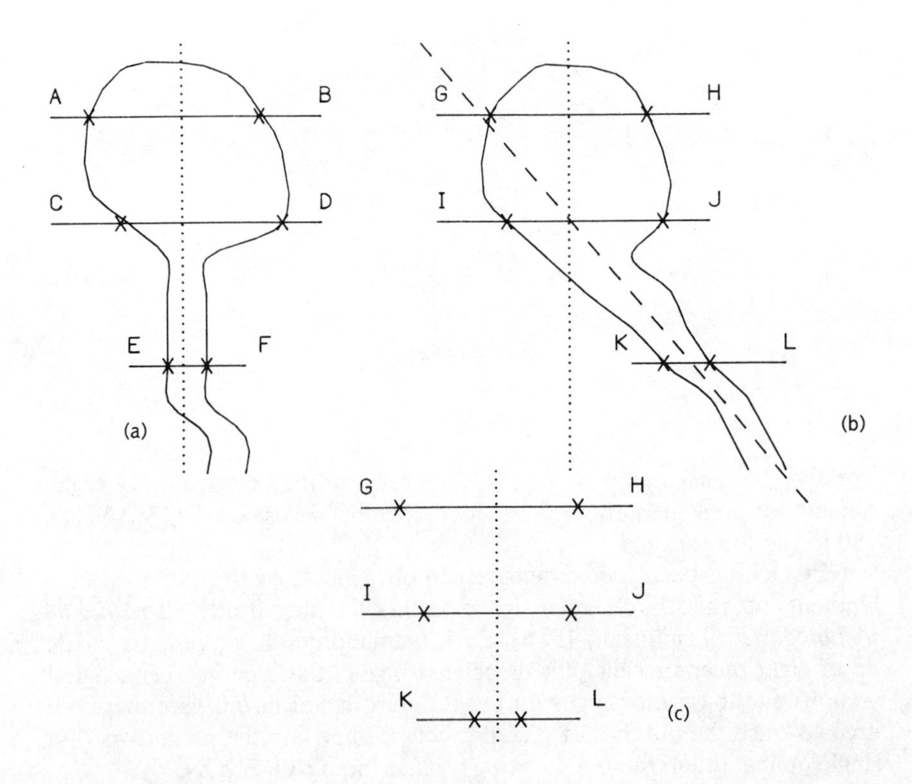

main display program, but extra 'copy' sections may be inserted to act as spacers to reproportion the picture as necessary. The system in its present form is of most use in morphological and developmental studies to look at spatial relationships, although more sophisticated analysis of quantitative aspects of morphology (as described by Moens and Moens, 1981) could be included in the main drawing program if required.

8.8 OTHER THREE-DIMENSIONAL RECONSTRUCTION PROGRAMS

Many other three-dimensional reconstruction programs have appeared in the biological literature in the last few years. Shepherd *et al.* (1984) describe a reconstruction system used to study oesophageal structure in a nematode. This system shares many of the general features of RECON, namely a data entry program, SSPROF, section alignment program, SSALIN, and a reconstruction and display program, SSRCON. These programs have been describd by Perkins and Green (1982). The SSPROF program includes a facility for entering general information about each series of sections, namely:

(1) Title.
(2) Magnification factor.
(3) Spacing between sections.
(4) Number of different categories (items like cells or organelles).
(5) Names to identify the categories.

The program is controlled by a menu on the upper edge of the tablet acting as input device. The available options are:

(1) Trace a closed outline.
(2) Trace an open outline.
(3) Enter a set of single points.
(4) Select next category.
(5) Select previous category.
(6) Select next section.
(7) Select previous section.
(8) Delete all of current section.
(9) Delete current category of current section.
(10) Insert a new section between the previous section and the current section.
(11) Renumber all sections to close any gaps.

Unlike RECON, SSPROF allows the user to observe data directly on a CRT as it is read into the computer. This allows rapid correction of data errors.

The SSRCON program offers a similar range of options as the RECON display routines, namely rotation, and section and category removal. SSRCON allows display of stereo pairs but does not include facility for hidden line removal. Representative output from SSRCON is shown in Plate 3. SSRCON has been used in a variety of different biological applications, for example formation of neural connections (Keating, 1980); study of axonal structures (Fawcett, 1981) and the assessment of lysosomal movement in macrophage cells (D'Arcy Hart *et al.*, 1983).

You will find many other examples of reconstruction systems listed in the bibliography below, and the range of applications is very broad. Nierzwicki-Bauer, Balkwill and Stevens (1983) used a similar system to study the three-dimensional ultrastructure of blue-green algal cells. A central component of this system is the ability to eliminate unwanted or unimportant cell structures, and the program was used to look at the intracellular locations of specialised inclusion bodies. Wong *et al.* (1983) present a general and simple computer reconstruction system suitable for implementation on small microcomputers (such as the Apple II series), and a number of authors discuss reconstruction of neural tree structures (for example, Zsuppan, 1984; Sasaki-Sherrington, Jacobs and Stevens, 1984; Sims and Macagno, 1985).

8.9 REFERENCES AND BIBLIOGRAPHY

The references cited here are by no means a complete list of all computer-aided reconstruction techniques in biology. They do, however, represent a fairly broad spectrum of the methodologies in use to date.

Balkwill, D.L., Stevens, S.E.Jr, Nierzwicki-Bauer, S.A. 'Use of Computer-aided Reconstructions and High-voltage Electron Microscopy to Examine Microbial three-dimensional Architecture. *Biotechniques*, 2 (4) (1984), 242–51

Burdett, I.D.J., Higgins, M.L. 'Study of Pole Assembly in *Bacillus subtilis* by Computer Reconstruction of Septal Growth Zones Seen in Central Longitudinal Thin Sections of Cells', *Journal of Bacteriology*, 133 (1970), 959–71

Capowski, J.J. 'Computer Aided Reconstruction of Neuron Trees from Several Serial Sections', *Computers in Biomedical Research*, 10 (1977), 617–30

D'Arcy Hart, P., Young, M.R., Jordan, M.M., Perkins, W.J. and Geisow, M. 'Chemical Inhibitors of Phagosome-lysosome Fusion in Cultured Macrophages also Inhibit Saltatory Lysosomal Movements', *Journal of Experimental Medicine*, 158 (1983), 477–92

Fairen, A., Peters, A. and Saldanha, J. 'A New Procedure for Examining Golgi-impregnated Neurons by Light and Electron Microscopy', *Journal of Neurocytology*, 6 (1977), 311–37

Fawcett, J.W. 'How Axons Grow Down the *Xenopus* Optic Nerve,' *Journal of Embryology and Experimental Morphology*, 65 (1981), 219–33

Foote, S.L., Loughlin, S.E., Cohen, P.S., Bloom, F.E. and Livingston, R.D. 'Accurate Three-dimensional Reconstruction of Neuronal Distributions in Brain Reconstruction of the Rat Nucleus Locus Coeruleus. *Journal of Neuroscientific Methods*, 3 (2) (1980), 159–74

Fuchs, H., Kedem, Z.M. and Uselton, S.P. 'Optimal Surface Resolution from Planar Contours', *Communications of the ACM*, 20 (10) (1977), 693–702

Green, R.J., Perkins, W.J., Piper, E.A. and Stenning, B.F. 'The Transfer of Selected Image Data to a Computer Using a Conductive Tablet', *Journal of Biomedical Engineering*, 1 (4) (1979) 240–6

Jordan, M.M., Perkins, W.J., Young, M.R. and D'Arcy Hart, P. 'Computer Graphics and Numerical Techniques in the Measurement of Movements within Cells', *Medical and Biological Engineering and Computing*, 23 (1985), 48–54

Keating, M.J. 'The Development of Neuronal Connections in the Retinal System of Lower Vertebrates', in C. di Benedetta (ed.) *Multidisciplinary Approach to Brain Development*, pp. 119–24 (North-Holland, Amsterdam, 1980)

Krieg, G., Cole, T. Deppe, U., Schierenberg, E., Schmitt, D., Yoder, B. and von Ehrenstein, G. 'The Cellular Anatomy of Embryos of the Nematode *Caenorhabditis elegans*: Analysis of Serial Section Electron Micrographs', *Developmental Biology*, 65 (1) (19789), 193–215

Martin, K.A.C. 'Neuronal Circuits in Cat Striate Cortex', in E.G. Jones and A. Peters (eds) *Cerebral Cortex*, Vol. 2 Chapter 9. (Academic Press, New York, 1984)

—, and Whitteridge, D. 'Form, Function and Intracortical Projections of Spiny Neurons in the Striate Visual Cortex of the Cat', *Journal of Physiology*, 353 (1984), 463–504

Mazziotta, J.C. and Hamilton, B.L. 'Three-dimensional Computer Reconstruction and Display of Neuronal Structure', *Computer Biology and Medicine*, 7 (4) (1977), 265–79

McIntòsh, J.R., McDonald, K.L., Edwards, M.K. and Ross, B.M. 'Three-dimensional Structure of the Central Mitotic Spindle of *Diatoma vulgare*, *Journal of Cell Biology* 83 (1979), 428–42

Moens, P.B. and Moens, T. 'Computer Measurements and Graphics of Three-dimensional Cellular Ultrastructure', *Journal of Ultrastructural Research*, 75 (2) (1981), 131–41

Nagurka, M.L. and Hayes, W.C. 'An Interactive Graphics Package for Calculating Cross-sectional Properties of Complex Shapes', *Journal of Biomechanics*, 13 (1) (1980), 59–64

Nierzwicki-Bauer, S.A., Balkwill, D.L. and Stevens, S.E.Jr 'Use of a Computer-aided Reconstruction System to Examine the Three-dimensional Architecture of *Cyanobacteria*', *Journal of Ultrastructural Research*, 84 (1) (1983), 73–82

Perkins, W.J. (1983) Interactive Computer Models of Biological Systems, in Geisow and Barrett (eds) *Computing in Biological Science* (Elsevier Biomedical Press, Amsterdam, 1983)

— and Green, J. (1982) 'Three-dimensional Reconstruction of Biological Sections', *Journal of Experimental Medicine*, 158 (1982), 477–92

—, Barrett, A.N. Green, J. Reynolds, D. 'A System for the Three-dimensional Construction, Manipulation and Display of Microbiological Models', *Journal of Biomedical Engineering*, 1 (1) (1979) 22-32

Sasaki-Sherrington, S.E., Jacobs, J.R. and Stevens, J.K. 'Intracellular Control of Axial Shape in Non-uniform Neurites. A serial Electron Microscopic Analysis of Organelles and Micro-tubule in A-I and A-II Retinal Amacrine Neurites', *Journal of Cell Biology*, 98(4) (1984), 1279–90

Schierenberg, E., Carlson, C. and Sidio, W. 'Cellular Development of a Nematode: Three-dimensional Computer Reconstruction of Living Embryos', *Wilhelm Roux's Archives of Developmental Biology*, 194 (2) (1984) 61–8

Shantz, M.J. and McCann, G.D. 'Computational Morphology: Three-dimensional Computer Graphics for Electron Microscopy', *IEEE Transactions of Biomedical Engineering*, BME-25 (1) (1978), 99-103

Shepherd, A.M., Perkins, W.J., Green, R.J. and Clark, S.A. 'A Study of Esophageal Structure in a Nematode *Aphelenchoides blastophtorus* Using Computer Graphics Reconstruction from Serial Thin Sections', *Journal of Zoology*, 204 (1984), 271–88

Sims, S.J. and Macagno, E.R. 'Computer Reconstruction of all the Neurons in the Optic Ganglion of *Daphnia magna*', *Journal of Comparative Neurology*, 233 (1) (1985), 12–29

Stevens, B.J. 'Variation in Number and Volume of the Mitochondria in Yeast According to Growth Conditions: A Study Based on Serial Sectioning and Computer Graphics

Reconstruction', *Biologie Cellulaire* 28 (1) (1977), 37–56

Tank, P.W., Connelly, T.G. and Bookstein, F.L. 'Cellular Behaviour in the Anteroposterior Axis of the Regenerating Forelimb of the Axolotl *Ambystoma mexicanum*', *Developmental Biology*, 109 (1) (1985), 215–23

Tomlinson, A. 'Analysis of Retinal Development in the Fruitfly Drosophila Melanogaster', *Journal of Embryology and Experimental Morphology*, in press (1985)

Veen, A. and Peachey, L.D. 'TROTS: A Computer Graphics System for Three-dimensional Reconstruction from Serial Sections', *Computing and Graphics*, 2 (3) (1977), 135–50

Von Swieton, A. and de Hosson, J. 'Remarks on Algorithm 475', *ACM Transactions on Mathematical Software*, 5 (1979), 521–3

Wong, Y.-M.M., Thompson, R.P., Cobb, L. and Fitzharris, T.P. 'Computer Reconstruction of Serial Sections', *Computer and Biomedical Research*, 16 (6) (1983), 580-6

Wright, T. 'Visible Surface Plotting Program', *Communications of the ACM*, 17 (1974), 152–6

Zsuppan, F. (1984) 'A New Approach to Merging Neuronal Tree Segments Traced from Serial Sections', *Journal of Neuroscientific Methods*, 10 (3) (1984), 199–204

9 Image Capture and Image Analysis

9.1 BIOLOGICAL IMAGES

We saw in the last chapter that computer graphics has been widely used in the reconstruction of sectioned biological material. A prerequisite of this type of analysis is that individual elements should be identifiable on the sections for digitization. Often, however, this kind of analysis is not feasible: the complexity of the data may preclude digitization altogether, for example, and in many cases the computer analysis is itself needed to work out the two-dimensional structure of the section.

The analysis of periodic images like muscle structure presents such an example, and the use of tomography to study structure by 'non-invasive' methods is a second example of a system that relies on more sophisticated image capture technology than the simple digitizer. In this short chapter we will look at several image analysis studies of interest to the biologist.

9.2 IMAGE CAPTURE DEVICES

Let us first clear up a point of terminology. We define image capture as the process by which an image, say a micrograph, is transferred to computer memory. The main difference between image capture and digitization is that the goal of the former is to transfer all the information from image to computer, whereas digitization usually takes the form of transfer of limited data to computer memory. Image analysis refers to the computer analysis performed on the image data in memory.

Image capture equipment based on video input is now widely used for the analysis of biological material. Unlike use of a digitizer, video input systems are usually supplied with software to at least transfer the image from source to computer, and the software will probably also have some manipulative routines to enable adjustment of the image in the computer. Two levels of system can be identified: the cheaper systems are general purpose video input devices with low to medium resolution (up to 512 × 512 pixels) designed to be coupled to personal microcomputers, and the software offered performs only limited manipulations. Examples are the Magic system (Heyden Datasystems, London) for use with the Apple Mackintosh computer, and the Digithurst Microsight systems (Digithurst Ltd, Royston, Herts, UK) used with a range of microcomputers including the IBM PC. More expensive microcomputer-based systems offer complete image analysis of two-dimensional data, including high resolution display

(1024 × 1024 pixels), full colour highlighting and quantitative analysis. The Joyce-Loebl Magiscan 2A is a system of this type.

The Digithurst Microsight 1 System (Figure 9.1) consists of a conventional TV camera, an interface between TV camera and computer, and a software package that controls the system. The software allows a number of operations to be performed, including the merging of images in memory to enable filtering, image mixing and superposition, together with a number of housekeeping operations like the saving and reloading of an image and dumping of the image to a dot matrix printer.

Single image capture takes around five seconds, and a 256 × 256 pixel resolution image with a 0-255 greyscale is generated. The image displayed on a CRT has only seven grey levels, but the software allows the separation of these levels, and specific density of the central grey level can be selected for the displayed image. Two types of video camera are commonly used in image capture work. The Digithurst system uses either VIDICON (for example, Ikagami ITC-40) or solid state (for example, Hitachi VK V2000E) cameras, the latter giving a more uniform signal across the field of view: VIDICON signals drop markedly for peripheral parts of the image, although these cameras are cheaper than solid state devices.

The Digithurst system can be connected to a variety of microcomputers

Figure 9.1: Image Capture System using the Digithurst Imaging System. From right to left can be seen a CCTV camera directed at a backlit electron micrograph negative, a monochrome monitor with the boxed interface placed on top, and a Sirius 1 microcomputer. The monitor is only used for setting up the image. The image displayed on screen is that of the M-band in frog sartorius muscle. Photograph kindly supplied by Pradeep Luther (Squire and Luther, 1985)

including the Apricot, Sirius and IBM PC. For simple applications involving fairly low memory requirements, image analysis will be performed on the microcomputer. Larger applications like three-dimensional reconstructions may require transfer of the image files to a mini or mainframe computer.

The Microsight I system will provide sufficient flexibility for many biological applications, but a more sophisticatd package, Microsight II, is also available. This system offers a computer-independent fast video framestore, image capture at 512 × 512 pixel resolution with 64 grey levels on screen (image capture at 40 milliseconds). The system also includes a command processor with a number of special functions that may be incorporated into applications programs. The functions are:

SETUP initializes and clears the frame store.

SNATCH reads one frame from the camera into the 512 × 512 frame store memory.

PUTPOINT(x,y,i) writes an intensity or grey scale value i from the computer to frame store pixel (x,y).

GETPOINT (x,y,i) reads an intensity value i from framestore pixel (x,y), into the computer.

READ (segment address) reads the whole image from the framestore into computer memory starting at the segment adress.

EXAMINE(x,y,i) reads the intensity value i from image location (x,y) within the frame data stored in computer memory.

DISPLAY transfers data from computer memory to the framestore for display on a local high resolution monitor; the display option includes a list of parameters for formatting, including window size and position, and for defining the type of processing.

9.3 ANALYSIS OF PERIODIC IMAGES

We will not attempt to review the vast field of image analysis in the present chapter (see for example Gonzales and Wintz, 1982), but an example of the use of image analysis in biology is offered by the analysis of periodic images.

The analysis of periodic images may be performed by the application of image averaging techniques using both Fourier transform and real space methods on images captured from electron micrographs. The aim of such analyses is to investigate the form of repeating units of biological structure, for example macromolecules, muscle, or proteins. Such image averaging techniques have been described by Klug and De Rosier (1966) and by Fraser and Millward (1970). Earlier work used optical diffraction and filtering or microdensitometer scanning to obtain the image to be processed. As microdensitometers are expensive (upwards of £30000), only a few research groups could afford to indulge in this type of study.

Devices like the Digithurst Microsight I system offer a cheaper solution to the image capture problem, and Squire and Luther (1985) describe the use of the Microsight I system to study the structure of muscle and collagen. The traditionally used Fourier transform methods (Brigham, 1974) are rather mathematical, and are not understood by the majority of biologists. Real space methods on the other hand offer a number of advantages including speed and simplicity, and Squire and Luther describe the rationale behind such methods. The following description is based on the method described by these authors.

Consider Figure 9.2, showing a two-dimensional array of repeating objects. The repeat direction is shown here parallel to the x axis. The repeating structures are situated x_d units apart along the x axis. As the objects need not be aligned along the y axis (unless the pattern is orthogonal), the y repeat distance will be y_d on the y axis, with a displacement dx on the x axis. It is often possible to define more than one repeat cell for a given pattern. An alternative cell with dimensions x'_d, y'_d in

Figure 9.2: Geometrical Description of Periodic Structures (see text for description of algorithm used). After Square and Luther (1985)

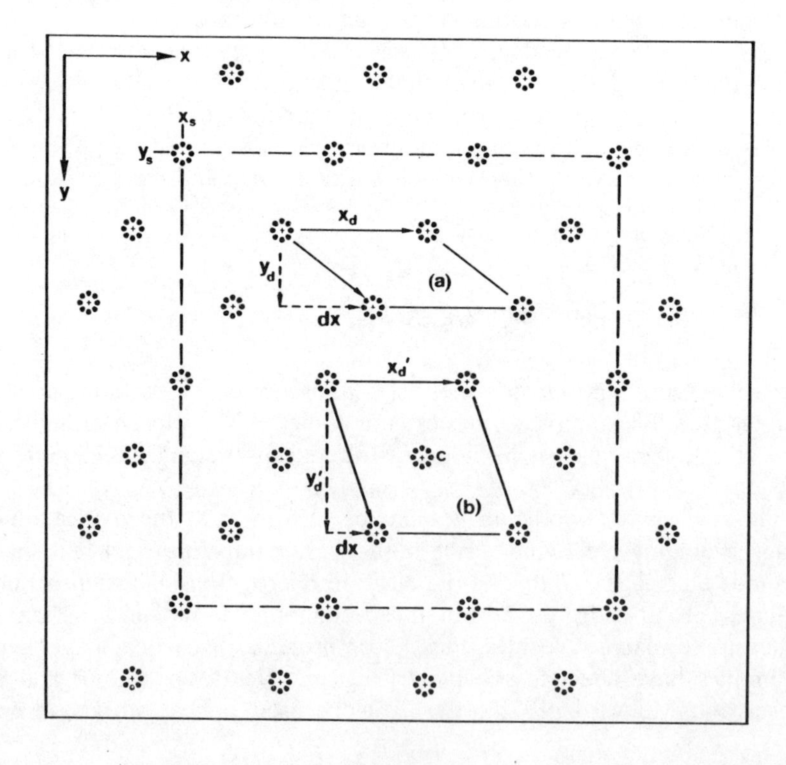

Figure 9.2 illustrates a second possibility. The advantage of defining multiple cells is that two independent average images of the repeating units are obtained, and the similarity or difference between these averages gives a built-in indication of the reliability of the image.

Once the repeat parameters have been set, averaging can be performed by:

(1) Choosing a start point of the array (for example, (x_s, y_s)).

(2) Stepping along x and y axes by a set number of points (for example M for x, N for y).

(3) Applying an averaging algorithm of the following form to each point

$$D(u,v) = \frac{1}{M,N} \sum_{m,n=0}^{M-1,N-1} d[(x_s + u + m \, . \, x_d + n \, . \, dx),(y_s + v + n. \, y_d)]$$

where

$D(u,v)$ = average density at the point (u,v) in a rectangular array $x_d y_d$
$d(x,y)$ = the stored density at point (x,y) in the original data array,

To construct the average image, the data in the average array $D(u,v)$ can then be translated to sequential cells of the whole array.

Figure 9.3 shows the result of applying the two-dimensional averaging procedure to a micrograph of the frog sartorius muscle Z band (Squire and Luther, 1985). Two different unit cells have been chosen (note that $dx = 0$ in both cases). The characteristic 'basketweave' pattern of the Z band is very clear in the averaged image.

9.4 THE JOYCE-LOEBL MAGISCAN

The Joyce-Loebl Magiscan (Figure 9.4) is based around a 16-bit central processor with its architecture optimised for image processing. This is a specialised custom designed processor offering advantages of speed and the ability to handle complex high resolution images; 128 Kbytes memory are present for program storage. An image store of 1024 × 1024 × 8 bits is present, giving 256 simultaneous grey scales or colours, expandable up to 16 bits (= 32768 colours/grey scales). A high quality Bosch TYK9A camera is used with a standard C-mount lens mounting so that a range of 35 mm type lenses may be fitted. Several different tube types are offered giving different resolutions (about 800 lines with the standard

Figure 9.3: A Composite Figure Showing an Example of Image Averaging Using a Real Space Image-Averaging method. (a) A transverse section of the frog sartorius muscle Z band; (b) the averaging function used; (c) the enhanced image showing the 'basketweave' pattern of cross connections on a square array of actin filaments. The unit cell size is shown by the dashed lines in (b). Micrograph provided by Pradeep Luther (Squire and Luther, 1985)

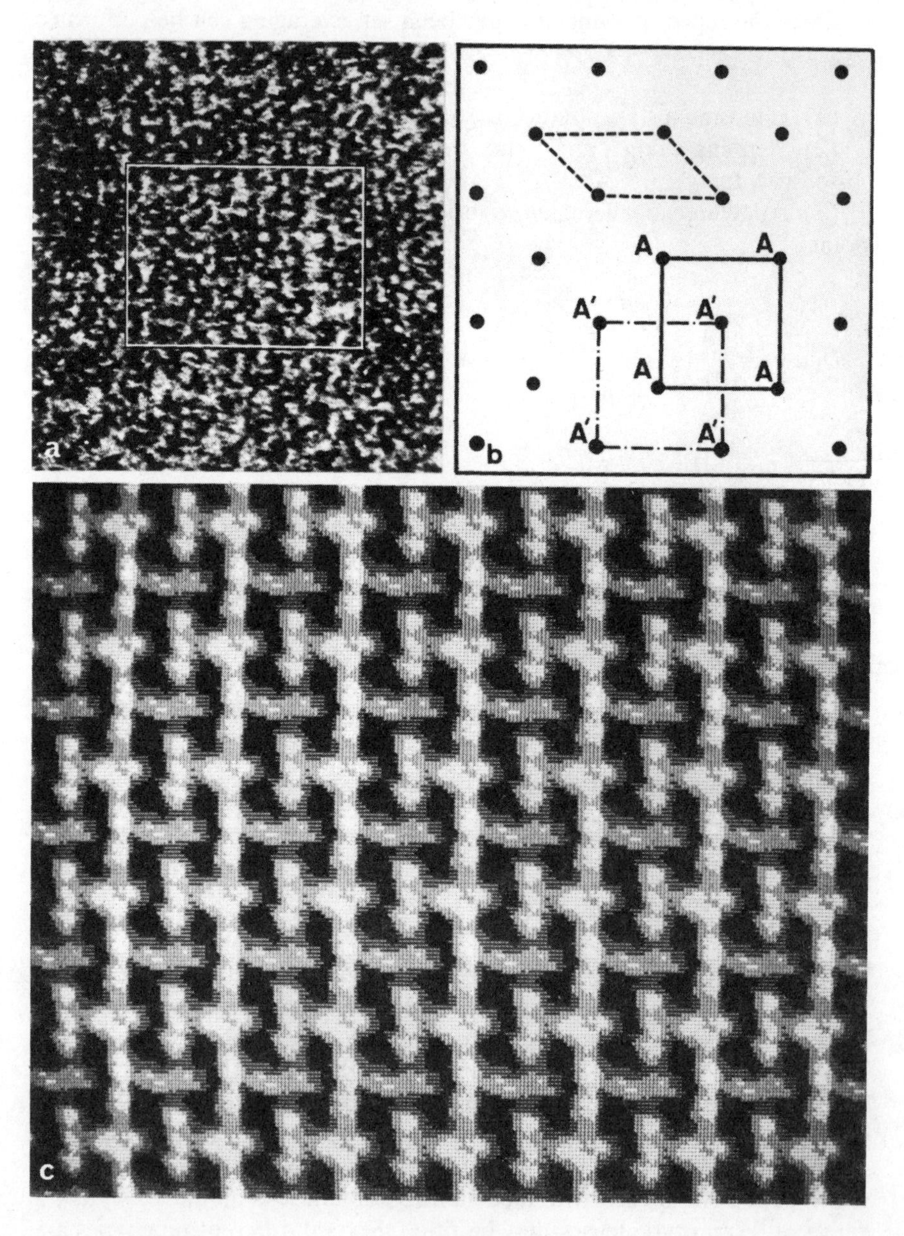

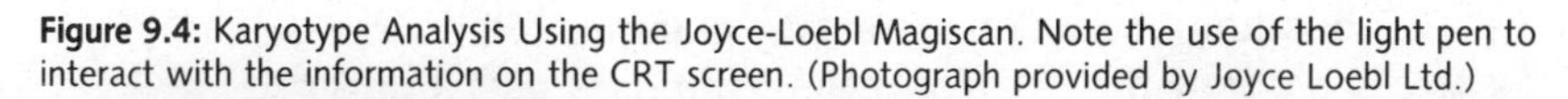

Figure 9.4: Karyotype Analysis Using the Joyce-Loebl Magiscan. Note the use of the light pen to interact with the information on the CRT screen. (Photograph provided by Joyce Loebl Ltd.)

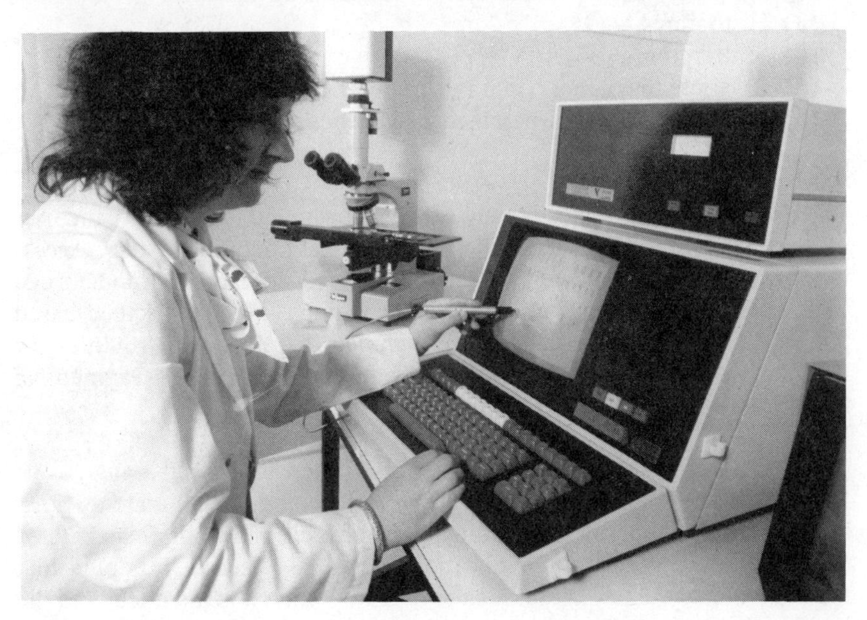

Chalnicon tube). The camera system can be adapted to fit most research microscopes, or alternatively it can be mounted in a stand and used with sketches, photographs or macroscopic objects.

'Menu'

Software for the Magiscan is organised in three packages. The first of these is called 'Menu', and allows the user to define environmental parameters like calibration and measurement units and frame parameters. 'Menu' accesses routines for separating features of interest from the background, including methods to cope with uneven illumination or poor contrast. Images can be edited to remove unwanted features or to connect broken objects. A light pen permits manual image editing. Another important inclusion is facility for coding of grey scales into colours: an extremely useful feature for highlighting differences between regions otherwise difficult to distinguish.

'Results'

A full range of measurements including counts, areas, lengths and optical densities can be obtained, and these are stored in data files to be read by the second package called 'Results'. This menu-driven package performs the following analyses:

(1) Tabulation of data (algebraically manipulated if required).
(2) Simple statistics (such as means, standard deviations).
(3) Histograms.
(4) Scattergrams.
(5) Bivariate histograms (that is, three-dimensional).
(6) Parametric and non-parametric hypothesis testing.

'Megamenu'

The third package is picturesquely called 'Megamenu', and allows the user to write Pascal programs using the UCSD Pascal operating system. Access to most of the routines used by 'Menu' is allowed, so that programmers can write special applications programs not included in 'Menu'. A screen-based text editor and Pascal compiler are included, together with routines for handling grey levels, boundary curvature and edge detection. Geometrical transformations are also included.

The Magiscan system is therefore a sophisticated tool for the analysis of histological material. It is relatively easy for the computer-illiterate research biologist to use, but the programmability offered allows the experienced user to produce specialised applications packages. There is no doubt that the system is extremely useful for the uses for which it is intended, and the limiting factor is really cost, especially as Magiscan is essentially a single user system.

9.5 RECONSTRUCTIONS FROM X-RAY DATA

In this section we will briefly discuss reconstructions based on tomography. These methods of 'non-invasive reconstruction' are widely used in the medical area, but have not been commonly employed in biological research, mainly because of the high equipment costs. This situation may change as techniques improve and hardware becomes more readily available.

Let us first consider the basic two-dimensional tomographic method. The history of these reconstructions began in the early 1920s, when several radiologists independently devised X-ray systems to visualise a two-dimensional section through a three-dimensional object. The method is shown in Figure 9.5. A photographic plate, P, is placed parallel to a movable X-ray source S, and between the two is positioned the object to be studied. The X-ray source and photographic plate are then moved in opposite directions at a constant speed. At one plane (C) parallel to the source and plate, points will project onto the same point in P, while points above and below C will be projected onto different points in P as it moves. So on the photographic plate, the density distribution at C will stand out, while the rest of the object will be blurred out (Gordon and Herman, 1974).

The major problem with the use of this 'linear tomography' method is that the output is complicated by interference from the density differences outside the plane of interest. A partial solution is to use 'transverse section scanning', in which all views are made in a single transverse plane (Figure 9.6). The density at each point is estimated by the sum of the total densities of all rays (the ray sum) through the point: this method is called summation.

If a number of scans of an object are used to reconstruct the final image, computer analysis is needed to process the large amounts of data

Figure 9.5: The Tomographic Method. *A* and *B* are two points in the object cross-section *C*. *X*1 and *X*2 are the positions of the X-ray source at two time intervals *t*1 and *t*2. *P* is the photographic plate. *A*1,*A*2 and *B*1, *B*2 are the positions of two fixed points on *P* at times *t*1 and *t*2. After Gordon and Herman (1974)

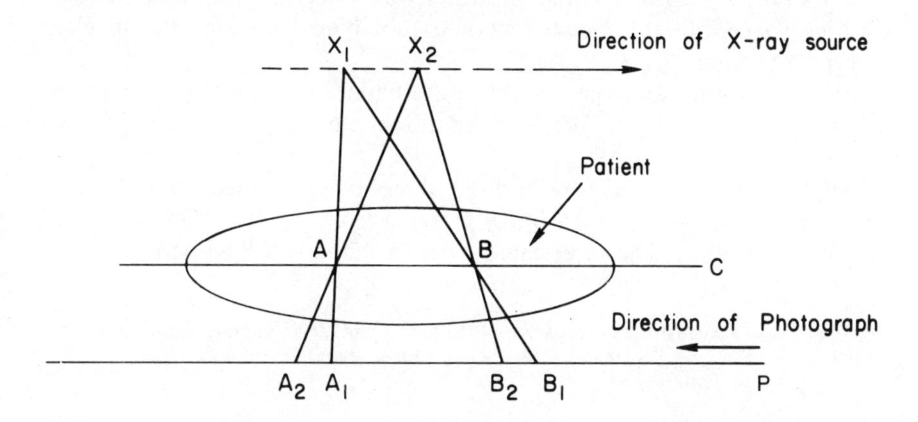

Figure 9.6: Reconstruction of a Plane of a Three-dimensional Object Using the Summation Method. The plane of scanning can be moved vertically to reconstruct sections of the whole three-dimensional object (after Kuhl and Edwards, 1963)

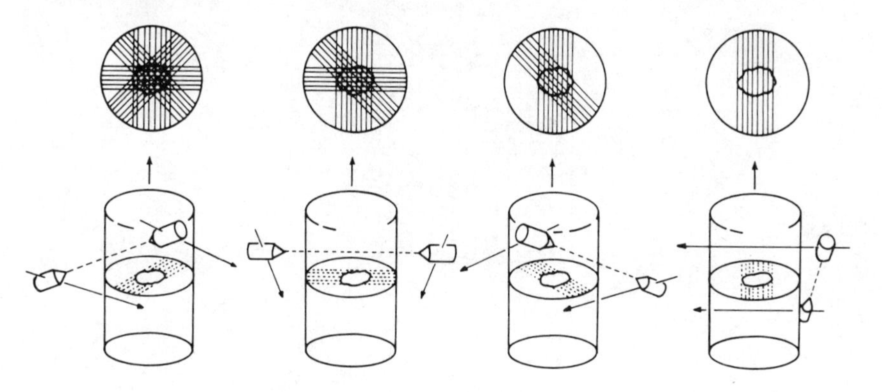

generated. The most widely used computer method for reconstructions of this type is called ART (short for Algebraic Reconstruction Technique; Gordon and Herman, 1974). ART is an iteration method that assigns an initial set of X-ray densities to the two-dimensional picture it is to reconstruct (for example, zeros). It then calculates the ray sum of each point along a one-dimensional projection, compares that ray sum with the existing density at that point in the computer, calculates the difference, and then divides it among all the pixels intersected by the ray. As the rays of one projection cross the rays of other projections, each computation affects the previous computations. This operation is then repeated for all the rays from all the projections until the picture is reconstructed (Figure 9.7).

Other algorithms besides ART have also been developed (reviewed in Gordon and Herman, 1974). These include:

(1) Fourier transformation, in which the data are transformed into Fourier space and the reconstruction is obtained by taking the inverse Fourier transform.

(2) Analytical solution of the differential equations (convolution method), a modified form of the summation method.

Three-dimensional display can be done by analysing stacked two-dimensional reconstructions (see for example Jaman, Gordon and Rangayyan, 1985). The three-dimensional volume will then be composed

Figure 9.7: Ray Geometry of a Cross-sectional Projection. From Jaman, Gordon and Rangayyan (1985)

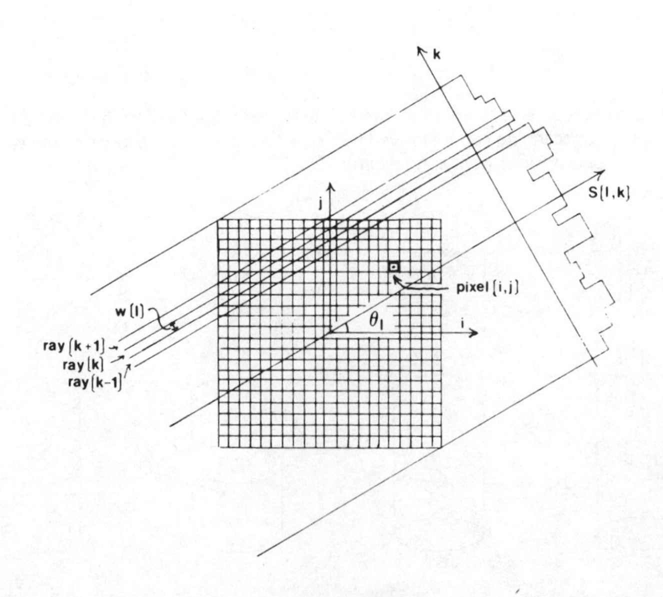

Figure 9.8: Projection Process for a Volume Image Showing the Projection of One Level of Voxels as an Example. From Jaman, Gordon and Rangayyan (1985)

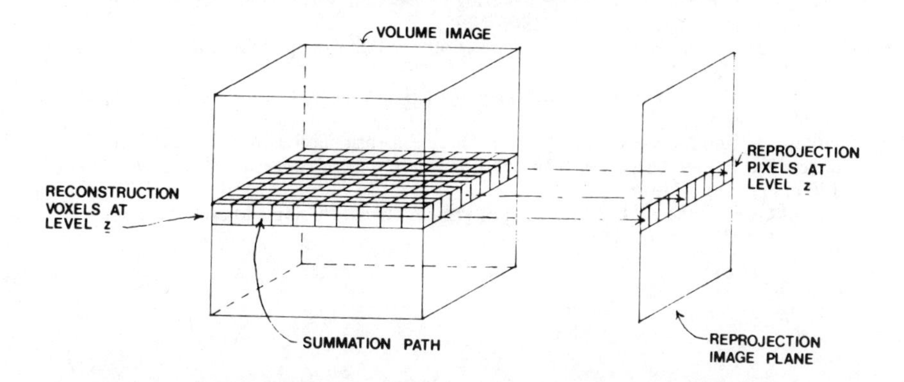

of small rectangular blocks (voxels), each of which represents the X-ray density of an element of the volume. The reprojection of the three-dimensional reconstruction onto a two-dimensional plane is performed numerically by adding the values of the voxels along the rays (Figure 9.8). Methods of 'peeling away' unwanted areas are also available, using what has been termed 'non-invasive numerical dissection'.

9.6 REFERENCES AND BIBLIOGRAPHY

Bender, R., Bellman, S.H. and Gordon, R. 'ART and the Ribosome: a Preliminary Report on the Three-dimensional Structure of Individual Ribosomes Determined by an Algebraic Reconstruction Technique', *Journal of Theoretical Biology*, 29 (1970), 483–7

Brigham, E.D. *The Fast Fourier Transform* (Prentice Hall, Englewood Cliffs, New Jersey, 1974)

Cook, P.N. 'Three-dimensional Reconstruction Algorithms. *Bulletin of the American Museum of Natural History*, 170 (1) (1981), 3–12

Crowther, R.A. 'Procedures for Three-dimensional Reconstruction of Spherical Viruses by Fourier Synthesis from Electron Micrographs', *Philosophical Transactions of the Royal Society of London Series B* 261 (1971), 221–30

De Rosier, D.J. and Moore, P.B. 'Reconstruction of Three-dimensional Images from Electron Micrographs of Structures with Helical Symmetry',. *Journal of Molecular Biology*, 52 (1970), 355–69

Dover, S.D., Elliott, A. and kernaghan, A.K. Three-dimensional Reconstruction from Images of Tilted Specimens of the Paramyosin Filament. *Journal of Microscopy (Oxford)* 122 (1) (1981), 23–34

Fraser, R.D.B. and Millward, G.R. (1970) 'Image Averaging by Optical Filtering', *Journal of Ultrastructural Research* (1970), 3

Gonzales, R.C. and Wintz, P. *Digital Image Processing*. (Addison-Wesley, Reading, Massachusetts, 1982)

Gordon, R. and Herman, G.T. 'Three-dimensional Reconstruction from Projections: a Review of Algorithms. *International Review Cytology*, 38 (1974), 121–51

—, Herman, G.T. and Johnson, S.A. 'Image Reconstruction from Projections', *Scientific American*, 233 (4) (1975), 56–68

Jaman, K.A., Gordon R. and Rangayyan, R.M. 'Display of 3D Anisotropic Images from Limited View Computed Tomograms', *Computer Vision, Graphics and Image Proceedings*, 30 (1985), 345–61

Klug, C. and De Rosier, D.J. 'Optical Filtering of Electron Micrographs: Reconstructions of One-sided Images', *Nature*, 212 (1966), 29

Kuhl, D.E. and Edwards, R.Q. 'Image Separation on Radioisotope Scanning', *Radiology*, 80 (4) (1963), 653-62

Ritman, E.L. *et al.* 'Three-dimensional Imaging of Heart, Lungs and Circulation', *Science* 201 (1980), 273–80

Schultheiss, R. and Mandelkow, E. (1983) 'Three-dimensional Reconstruction of Tubulin Sheets and Reinvestigation of Microtubule Surface Lattice', *Journal of Molecular Biology* 170 (2) (1983), 471–96

Squire, J.M. and Luther, P.K. 'Averaging of Periodic Images Using a Microcomputer', *Journal of Microscopy*, (1985)

10 Molecular Graphics

10.1 AN INTRODUCTION TO MOLECULAR GRAPHICS

The first thing that we should do in this chapter is to define precisely what we mean by molecular graphics. The term would seem to imply some form of graphical representation of molecular data, and indeed this is certainly true. However, there is much more to this area of research than just drawing pretty pictures. Let us consider a few examples to illustrate the varied applications of this powerful tool. Michael Connolly, when he was working at the University of California, San Francisco, developed a method of displaying the re-entrant surfaces of molecules on a very powerful graphics system (Connolly, 1984).

In essence, Connolly's system allowed the surface to be modelled as a field of coloured dots (see Plate 4), with the added advantage that the system was highly interactive. Thus, investigators could look at the shapes of active sites and visualise the 'lock-and-key' fit between molecular complexes, for example thyroxin–prealbumin, bpti-trypsin, haemoglobin, alpha and beta chains.

Professor Arthur Lesk of Fairleigh Dickinson University, New Jersey, developed a program for displaying the secondary structure of proteins as cylinders and ribbons. This form of representation is very popular when just an overview of the molecule is required, as opposed to the fine detail of an all atom model. This system is of great interest to those researchers involved in secondary structure techniques, structural comparisons, or secondary structure analysis. The new version of the Lesk system features high quality colour graphics (developed with K.D. Hardman; see Lesk, 1984; Chothia and Lesk, 1985).

Richard Feldman, of the National Institute of Health, Bethesda, Maryland, USA was and is one of the prime movers in the development of molecular graphics. His accomplishments are too numerous to do them justice in the limited space available here, (see, for example, Feldman, 1976; Feldman *et al.*, 1974). However, mention can be made of one study done with Michael Levitt (Feldman *et al.*, 1978). This involved the production of a film displaying a simulation of the motion of bovine pancreatic trypsin inhibitor.

The above examples are just the tip of the iceberg in the field of molecular graphics, for it would be very difficult to find a paper on molecular structure without at least one computer generated image. This is perhaps an exaggeration, but the point is that access to molecular graphics systems is quite common, at least at the larger universities. The graphics systems can range from simple ball-and-stick or space-filling (that is, intersecting spheres) programs run on a small micro to very large software packages for performing graphics and analysis which run on supercomputers, for example the Cray series, with expensive colour graphics devices.

10.2 COMPONENTS OF A MOLECULAR GRAPHICS SYSTEM

At this stage we consider the possibility that you may wish to avail yourself of some molecular graphics capability. The first step is, of course, to define your molecular problem. We will assume for the sake of argument that the problem is suited to graphical manipulation. What is the next step? Several choices are possible. You can use an existing system that you begged, borrowed or 'acquired'. Alternatively, you can write your own system. There is a third choice, but this is more difficult in practice: if you have access to the source code of an existing system, you can always try to modify it to accomplish your task. If you have ever tried to modify someone else's code you will understand the difficulties of this option. You are therefore left with the first two choices.

Having defined the problem and realised that there are two courses of action, you should next consider the hardware to be used. This is important if you are either choosing or developing your own system. We will assume that a major component of the system is graphics, but what type of graphics device should be employed? Recall that we examined a range of devices and their implications in Chapter 2. For most of us on humble budgets the graphics device chosen will probably be some form of medium resolution raster CRT. This would be quite adequate for most applications which do not require real-time transformations and extensive shading and lighting of the displayed molecule. For those on large budgets, an expensive vector device might be the best solution. This would allow the production and/or the manipulation in real-time of complex line drawings. Suffice it to say that the almost instantaneous response afforded by vector devices can be very useful, as in, for example, protein-ligand docking or any dynamic activity.

The next component of the system to be considered is that of interaction. That is to say, do you want to directly manipulate the system, say through a terminal, or alternatively to prepare some file of commands, process them and get your results. Clearly, this is an important consideration and your choice must reflect the types of problems and data you wish the system to manipulate.

The final component that we need to consider on this level is analysis. If the system you require just inputs some atomic coordinates, transforms them and then produces a displayed image, then you probably do not want to become involved with the analysis issue. If on the other hand you require, say, a system to aid you in conformational analysis, then you would need at least the ability to specify the bonds to rotate, the increment of rotation and the total number of degrees to rotate. There are of course other forms of analysis, for example energy calculations, for which due consideration must be given. Three examples are described below which meet the above criteria for

(1) a non-interactive general molecular system;
(2) an interactive and highly specific graphics system; and
(3) an interactive graphics and analysis system.

Before we discuss these example systems we look at perhaps the most important element of the system: the data.

10.3 MOLECULAR DATA

Given the best system in the world, if the data is incorrect or inappropriate then you will never achieve your desired results. Everybody knows this, yet we all make mistakes from time to time. Therefore, it is worth taking a little time to consider the nature of the data. As a working example let us look at the structure of proteins, although the basic methodologies will be the same for other types of molecules, for example nucleic acids or small molecules.

The primary level of protein organisation consists of a set of amino acids linked together into a long unbranched chain. The next level of organisation is the formation of the secondary structure, mainly via the formation of hydrogen bonds. That is, we may think of the secondary structure as a list of all three-dimensional regions that have ordered, locally symmetric backbone structures. Two of these ordered structures are the alpha helix and the beta sheet. The tertiary structure of a protein can be defined as the complete three-dimensional structure of one effectively indivisible unit. Finally we may consider the quaternary structure. This structure is the highest level, and is formed by the non-covalent association of independent tertiary structure units. The subunits of the quaternary structure may or may not be identical, and their subsequent arrangement in the quaternary structure may or may not be symmetric (Cantor and Schimmel, 1980). Thus, in a rather simplistic sense we have four basic levels in which to frame our enquiries. What kind of data would we encounter at these levels and how does molecular graphics aid us in the manipulation and analysis of this data?

Let us begin with the primary structure and look at its components. As we previously stated, this level consists of a linear sequence of amino acids. But these amino acids (recall that there are twenty of them) are in turn composed of various atoms in a highly organised system. So, even at this first level of enquiry we are faced with several choices of the level of analysis, that is, the amino acids or at the lower level, the atoms that comprise the amino acids. For any protein of even average size this can be a substantial amount of information to manipulate and understand, especially if one is faced with volumes of numbers. These numbers arise from the X-ray crystallographic analysis of the protein which leads to the three-dimensional coordinates of the individual atoms being determined.

What can we do with all this information? Clearly the strategy to be used is a function of the task at hand but molecular graphics can and does provide some powerful tools for the display and manipulation of this data. On a purely visual level we could decide to first display the backbone of the protein as defined by the positions of the alpha-carbons of the individual amino acids. Use of the alpha-carbons as representative atoms would clearly show the course of the molecule through three-dimensional space and would also illustrate any secondary structural elements such as helices, sheets and turns. An example of this form of drawing is given in Figure 10.1(a), where the individual alpha-carbons are represented by spheres. There are of course several other methods available to illustrate the same information. For example, we might eliminate the spheres and draw the whole backbone as a 'stick' figure, with each alpha-carbon represented by the bend between straight line segments. This could lead to difficult problems in perception as a reasonable-sized molecule would tend to look like a plate of spaghetti, due to lack of either perspective or three-dimensionality. We could extend this representation down to the individual atom level. Stick figures at this level cause even worse problems, except for the study of one or two amino acids.

The representation of the tertiary and quaternary structures is an extension of the primary and secondary structures, albeit made more complex by virtue of the addition of more information. While it might be useful to display the all atom representation of a quaternary structure from the perspective of complete molecular detail, its usefulness must be in some doubt. The sheet visual complexity of say, the quaternary structure of haemoglobin is overwhelming with an all-atom display. What would be of interest, and more feasible might be to look at the interfaces of the chains in this all atom detail. While complex, it would certainly be meaningful when viewed from the appropriate angles. Anderson and Cygler (1985) have used molecular graphics systems to predict the tertiary structure of the bacteriophage 434 Cro repressor. Both vector and space-filling representations have been used, the latter allowing the examination of the protein surface and the atoms exposed on this surface. Parts of the structure can also be removed to facilitate examination of internal regions for packing defects or inappropriately placed side-chains. By building models of the protein/DNA complex, Anderson and co-workers were able to define possible contacts between proteins and DNA (Plate 5). The

Figure 10.1: Two examples of an Alpha-carbon Representation of Proteins using PLUTO. (a) Lysozyme (Ball-and-stick representation with shading); (b) Cro (Ribbon drawing)

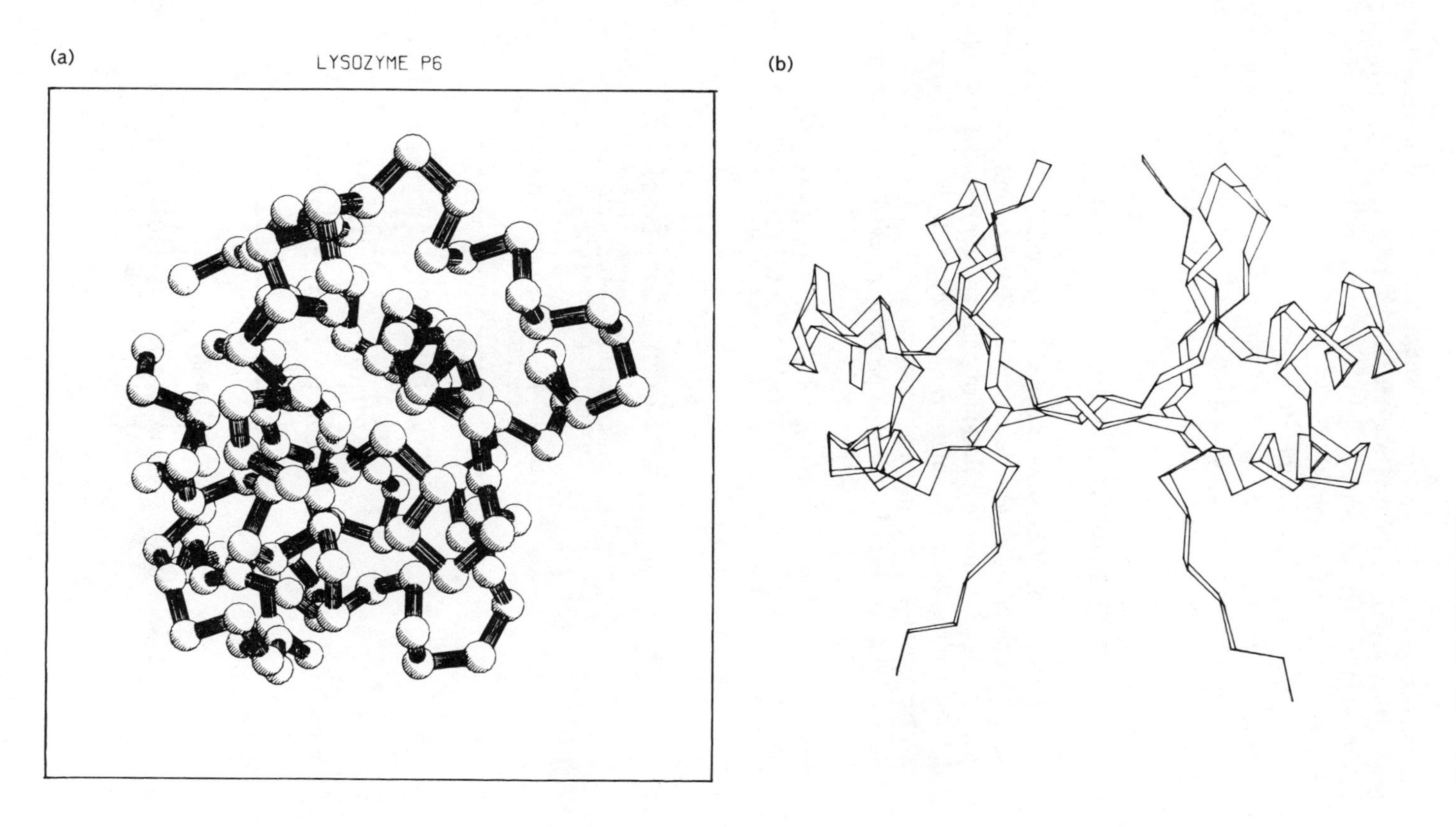

models suggested interactions between the amino acide side-chains and particular bases within the operator DNA (Figure 10.2); however, the three-dimensional structure of the Cro complex with DNA will have to be determined to test the model.

Molecular graphics can therefore be used as a powerful tool to aid the researcher in his or her task of understanding molecular structure and function. The choice of representation is entirely a function of the nature of the work and the display systems and software available.

10.4 EXAMPLES OF MOLECULAR GRAPHICS PACKAGES

We will begin our discussion of molecular graphic packages by considering a small space-filling system developed by Zientara and Nagy (1983). In a space-filling system the selected atoms are represented by spheres of defined radius. When the image is projected on to the viewing plane, the resultant display is one of interlocking circles. In essence the Zientara and Nagy system consists of a versatile program for displaying protein or poly-

Figure 10.2: Model of the Interaction of Amino Acid Side-chains with the Base Sequences of the Operators in the Right Operator Region of Bacteriophage 434. The interactions were suggested by a graphics model of the 434 Cro complex. (After Anderson and Cygler, 1985)

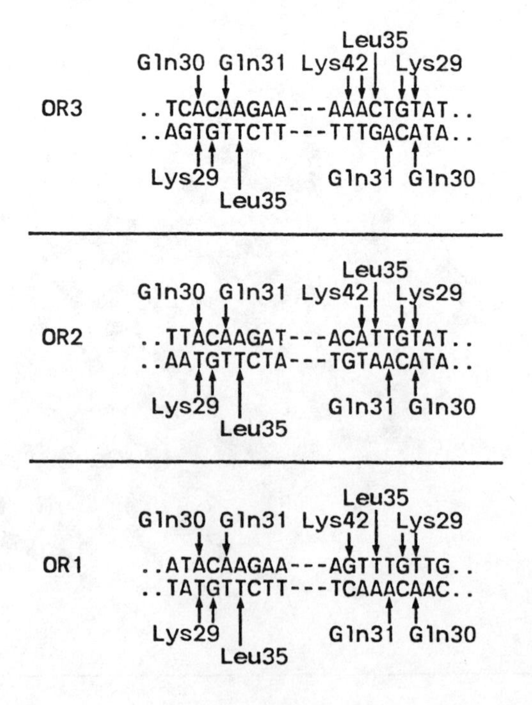

peptide molecular structures. The input consists of processed non-hydrogen atomic coordinates in standard protein databank (PDB) format. The system is interactive and after a short preliminary section, described below, produces a two-dimensional projection of the space-filling molecule. The interactive inputs consist of user-selected viewing angles, preferences in residue shading, the choice of drawing only backbone (that is, N-Ca-C) atoms, and the possible specification of a subset of residues for representation. The operational advantages cited by the authors are the amenability of the Fortran code to small systems and the ability to produce reasonable results on low cost plotters.

The overall program strategy is to provide storage for the input data, to reduce and transform the data set as requested by the user, and finally to display in an efficient manner the required structure. This is illustrated graphically in Figure 10.3 (a,b,c). From this we see that the protein databank input consists of ATOM, CONECT and HETATM records. Once this input data has been processed, the remaining components of the main program are concerned with the user's choice of molecular structure elements for drawing and the visual mode of that representation. The specification of the elements necessitate that the user interactively makes five major decisions at run time. These may be described as follows:

(1) The user decides whether a small subset of amino acid residues is chosen for display or alternatively whether all input data will be used.

(2) Given a complete set (or subset) of residues, the user next decides if all the atoms comprising both the backbone and side-chains or only those atoms within the backbone are to be displayed. If all of the atoms in a set of residues are to be displayed, the user must next decide if these atoms are to be visualised in their entirety, that is neglecting the visual interference of the remainder of the protein, or as they would appear in the complete molecule, with hidden lines removed.

(3) Select rotation parameters for viewing amino acids.

(4) A choice of scale is given with respect to the full molecule or the subset of amino acids.

(5) The specification of amino acid shading.

The program also provides for the selection of several residue subset categories which facilitate the rapid selection of specific amino acid residues for representation. Currently these include: (a) residue names; (b) charge group types; (c) structural types; (d) residue number; (e) all residues; (f) residues with specific solvation characteristics; (g) surface representation of specific residues. As an example, say we were interested in displaying only the alanines. We could choose category (a) and input the three-letter code ALA. This would produce a representation of only the alanines in the molecule.

Figure 10.3: Flowcharts of the Space-filling Molecular Drawing Package Devised by Zientara and Nagy (1983). (a) Main Program to set up data files; note links to (b), transformation routines, linking in turn to (c) output routines

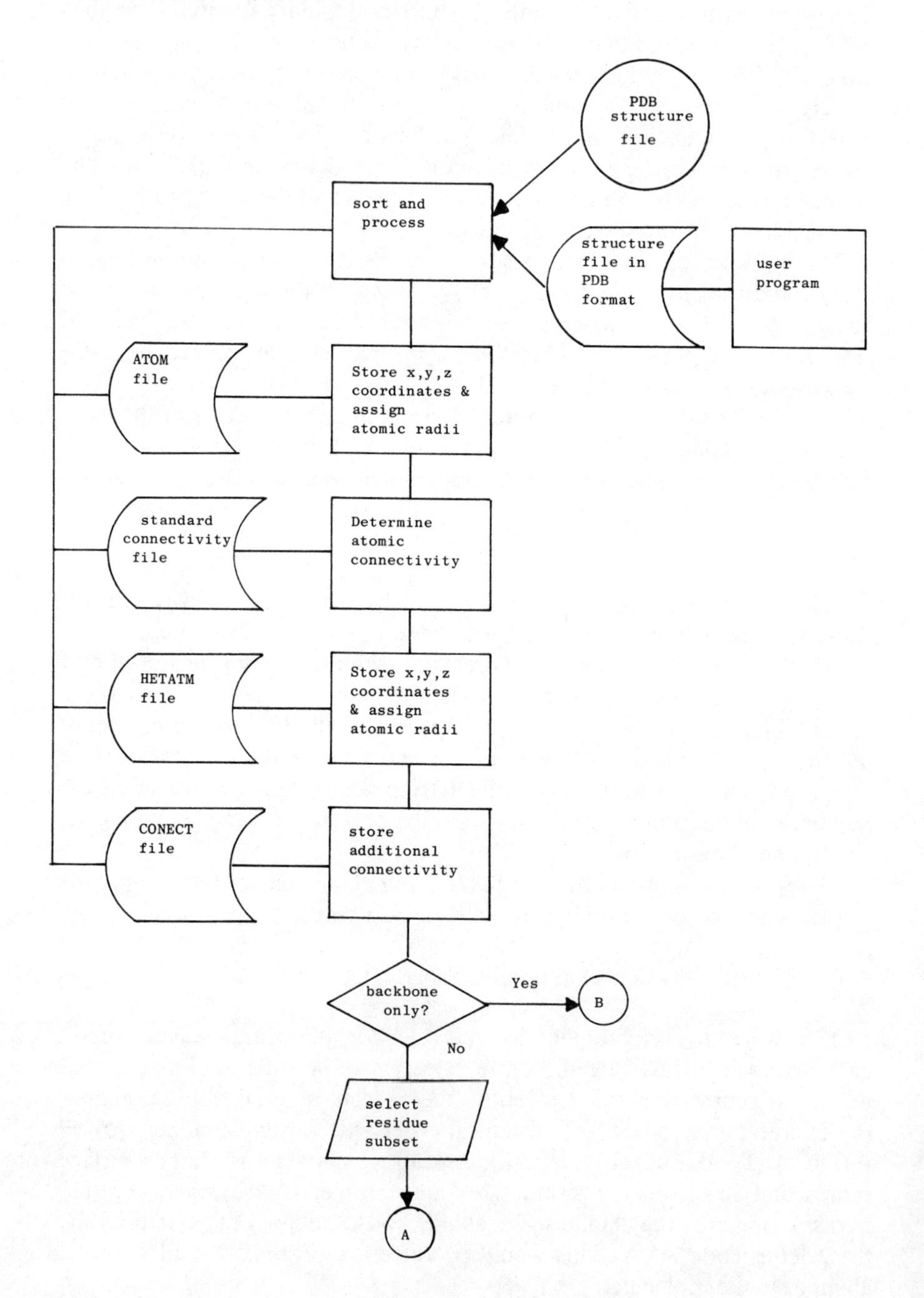

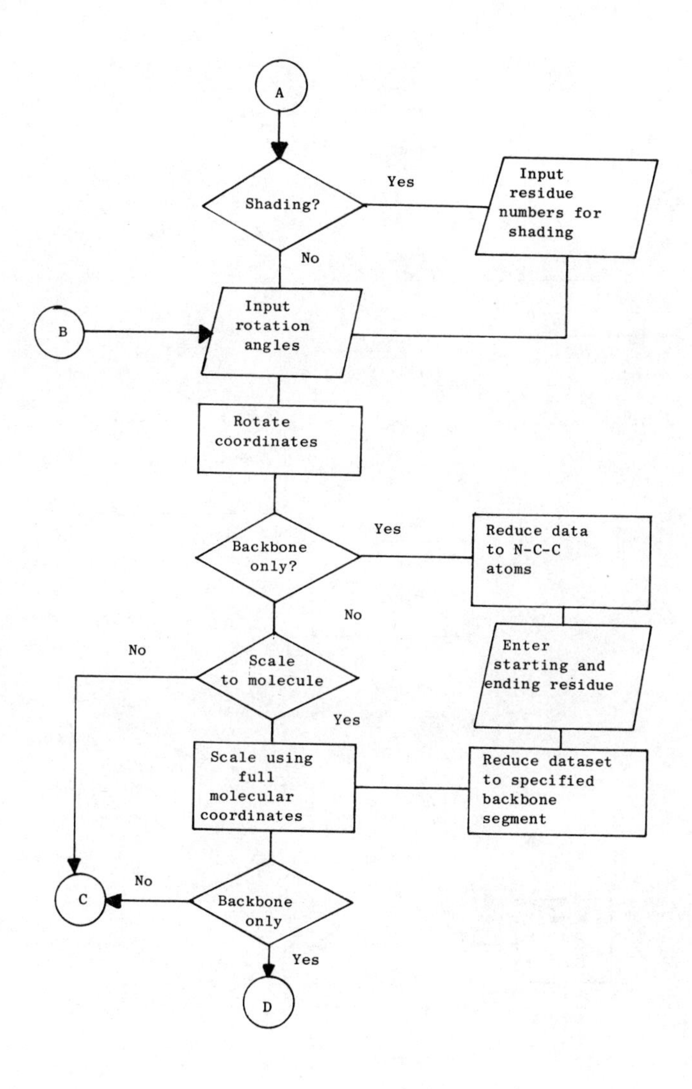
A
Shading?
Yes
Input residue numbers for shading
No
B
Input rotation angles
Rotate coordinates
Backbone only?
Yes
Reduce data to N-C-C atoms
No
Enter starting and ending residue
No
Scale to molecule
Yes
Scale using full molecular coordinates
Reduce dataset to specified backbone segment
No
C
Backbone only
Yes
D

Figure 10.3 continued

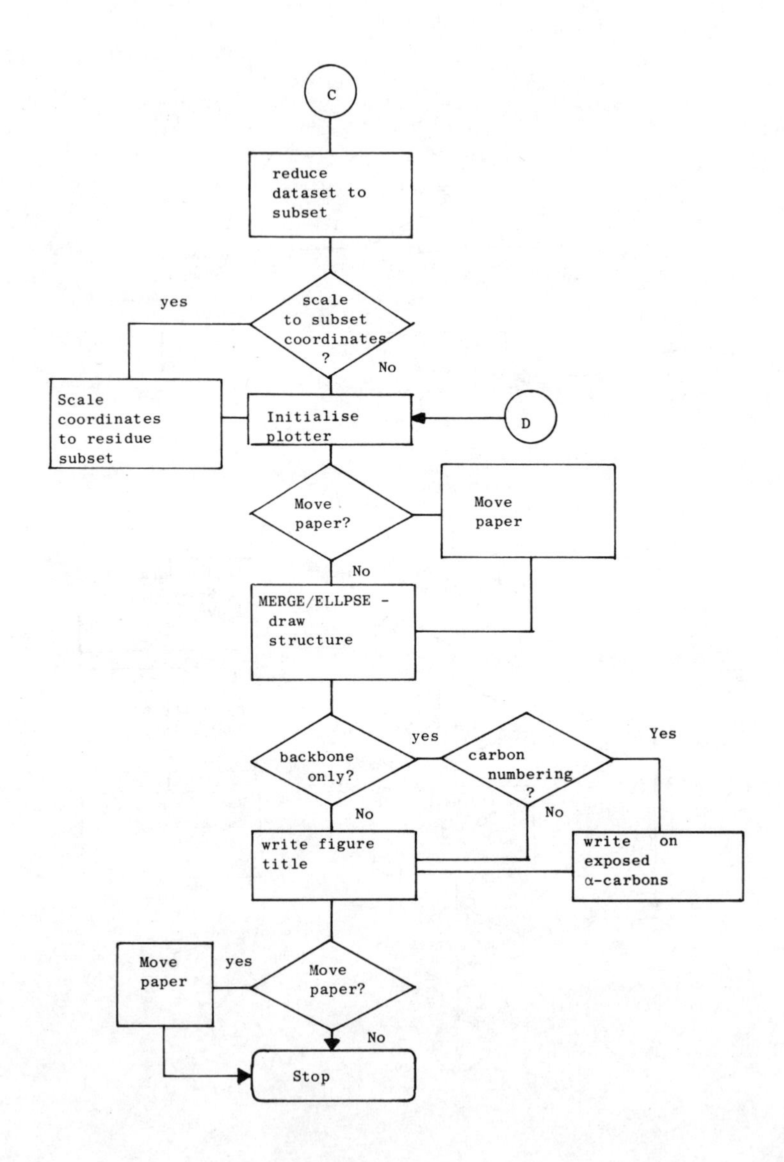

After the choice of molecular structure to be displayed, all calculations involving structural rotation and scaling are performed. These operations are carried out within the main program. Next the plotting routine, MERGE, is called. This subroutine performs all calculations and plotting which produce the interpenetrating sphere or space-filling representation of a molecular structure. Examples of this are given in Figure 10.4.

In the previous system we were limited to a space-filling representation of our molecular structure. We will now discuss a more general non-interactive system developed by Sam Motherwell at the University Chemical Laboratory, Cambridge. The system, PLUTO, was written for small molecules and was modified for proteins by Eleanor Dodson and for map plotting by Phil Evans.

PLUTO is described as a flexible program for plotting molecular and electron density maps, either separately or together. While the system is not interactive, there are considerable options within both these areas. For example, with respect to molecular drawings the options would include: (1) displaying the molecule as sticks; (2) ball-and-sticks; (3) space-filling; (4) parallel projection; (5) perspective; or (6) stereo pairs with perspective (either side by side or red/green). The map option would include (1) plotting single section; or (2) stacks in mono or stereo; and (3) maps may be plotted together.

We shall now look at some of the details of the system but confine ourselves to proteins. The first item that we consider is the choice of coordinate system. The program supports four main systems but only two are relevant to molecules. The coordinates of the molecule are normally given in the orthogonal form. These may be converted to fractional cell coordinates using the cell dimensions.

All commands in PLUTO are via control cards which, among other items, contain control key words. In essence, the control cards are read to set up the various facilities required. For example, after execution of the control INPUT and its associated cards, the coordinates of the atoms are placed into storage arrays. The JOIN command establishes the connection between the atoms, and the radii of the various atoms are specified by the RADII ATOM card. The PLOT command produces the output file, which is device-independent, for subsequent processing and display. The use of the term control card arises for an historical reason. As the system is non-interactive, PLUTO jobs could be run in batch mode, which in the old days meant that you submitted decks of cards. In the current system the physical card has been replaced with the notion of a record in a file.

Without going into all the detail of the control set, we can illustrate some of the commands with the following example.

Figure 10.4: (a) Residues 126–153 of Sperm Whale Myoglobin; Only the Backbone Atoms Have Been Represented. (b) Sperm whale myoglobin with residues 1–19 shaded. From Zientara and Nagy (1983)

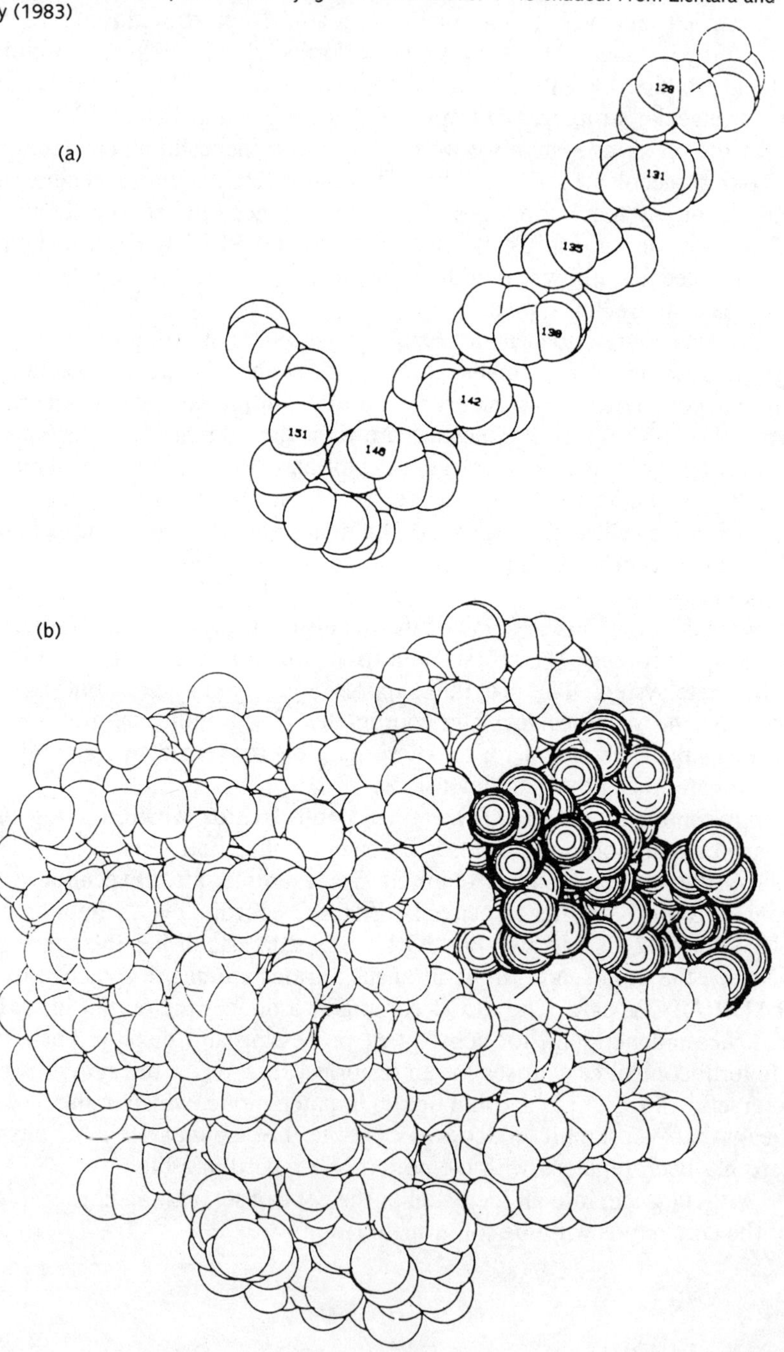

```
TITLE CRO REPRESSOR CRO1
CELL 1.   1.   1.
SYMM X,Y,Z
INPUT BROOK CAS
BOND 1 0.5 40
RESIDU SELECT 2 TO 67
RESIDU BOND 1 ALL
LINK CA CA 4.1
SOLID SHADE 120,−45
MONO
JOIN RADII RESIDU CA 2.0 CA 2.0 1
RADII ATOMS CA 1.
RADII BONDS .55 2
THICK 3
SIZE 200 CHAR 4
FONT 0
VIEW X
PLOT
```

A brief description of some of the above commands is as follows:

INPUT BROOK CAS: The input data is in Brookhaven format and we only want the alpha-carbons.
BOND 1 0.5 40: This specifies the characteristics of a bond of type 1. In this case the cylindrical radius is 0.5 mm and the number of lines defining the bond is 40.
RESIDU SELECT 2 to 67: The residues that we wish to consider are number 2 through to 67.
RESIDU BOND 1 ALL: This card states that all the bonds between the chosen residues are of type 1.
LINK CA CA 4.1: This card directs the system to link all the alpha-carbons that are within 4.1 angstroms (Å) of each other.
SOLID SHADE 120,−45: The alpha-carbon atoms are to be drawn with hidden line removal activated and the spheres are to be shaded. The 120 and −45 refer to the two angles used in the specification of the light source.
MONO: This is a monocular drawing as opposed to stereo.
JOIN RADII RESIDU CA 2.0 CA 2.0 1: This card controls the manner in which atoms are linked. In this case the keyword RADII limits the search to the sum of the bonding radii. The keyword RESIDU limits the distance search to within residues.
RADII ATOMS CA 1: This set the radii of all alpha-carbon atoms to 1 Å.
VIEW X: View along the x axis towards the origin.

The results of running this file and plotting the resultant data are given in Figure 10.5. We also illustrate an all-atom representation from PLUTO in Figure 10.6, which shows a beta sheet from superoxide dismutase.

The final system that we shall consider in detail is MOLY — an interactive system for molecular analysis (Dyott *et al.*, 1980). The principal differences between MOLY and the previous two systems lies in the fact that MOLY not only allows for graphical manipulation but also has the

Figure 10.5: The Cro Repressor Drawn Using PLUTO

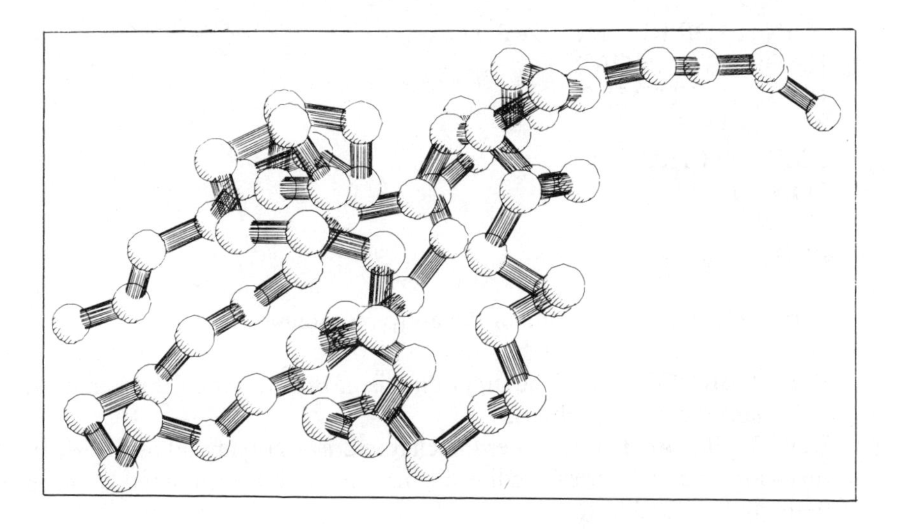

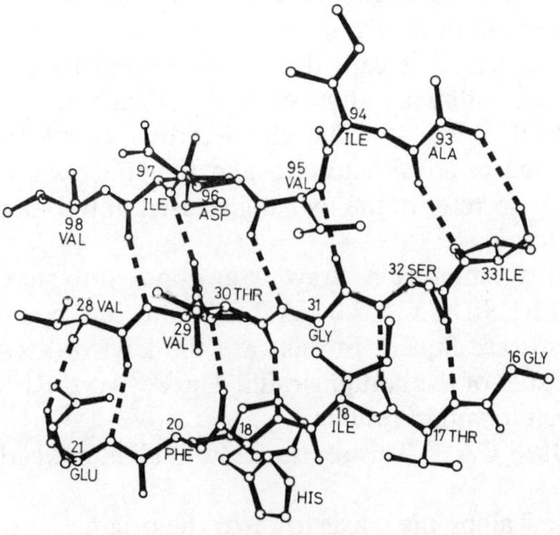

Figure 10.6: An All-atom Representation of Superoxide Dismutase Drawn Using PLUTO

ability to perform detailed conformational analysis. Indeed, a summary of the capabilities of MOLY would include: provision for the rapid construction and manipulation of three-dimensional, chemically correct molecular images; the facility to make comparisons between different molecules; as stated above, detailed conformational analysis; the estimation of a molecule's lipophilicity; and finally the provision of a limited description of a molecule's electronic properties.

The authors state that MOLY is an interactive computer graphics system with the following characteristics: (1) the system is quite large (about 800 Kbytes); (2) is modular, to facilitate additions and enhancements; (3) is command-driven; (4) contains extensive user prompting; and (5) is heavily dependent on graphics.

In general MOLY consists of a main section of code which acts as a driver to decipher the user's instructions and evoke the appropriate response. There are seven major modules that interact with this driver, whose primary functions may be described as follows:

(1) INPUT: enter molecules by drawing them in via a graphics terminal.

(2) MODEL: build a reasonable three-dimensional model of the molecule.

(3) CONFOR: perform conformational analysis.

(4) ANALYZ: prepare contour maps of energy and population as a function of rotation about one or two bonds.

(5) LOG P: calculate octanol/water partition coefficient.

(6) FORMAT: interface with batch programs for CNDO/2 and PCILO quantum mechanical calculations.

(7) COMPARE: using non-linear least-squares regression, map one molecule onto another.

The MOLY system incorporates a very friendly user interface for entering chemical structures for subsequent display and analysis. The actual input consists of drawing the structure on a terminal which is capable of graphics input. This is done in the same manner as one would do on a piece of paper. When a structure is entered, the screen is divided into two distinct areas: the drawing page (enclosed in a box), and a command menu as shown in Figure 10.7. This system coupled with the appropriate graphics device, allows the structure to be drawn or a command to be selected via the cursor control system, such as the joystick.

The normal input sequence would entail the following:

(1) clear the drawing area of any redundant structures;

(2) enter the DRAW mode and draw the molecular skeleton, while indicating the appropriate bond types (that is, 1 = single, 2 = double, 3 = triple, 4 = aromatic);

Figure 10.7: The Molecular Input Drawing Page Used in MOLY

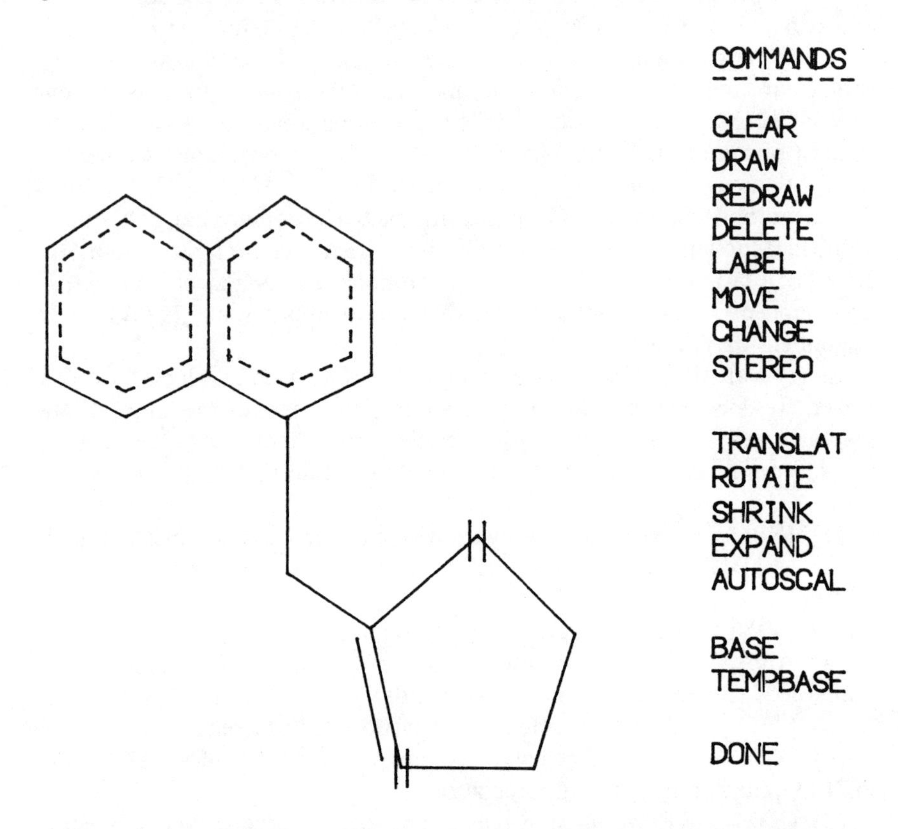

(3) enter the LABEL mode and specify the heteroatoms by pointing at them with the cursor and typing in the relevant chemical symbol (all 103 elements are recognised);

(4) if necessary, enter CHARGE mode to specify formal charges.

Hydrogens can be explicitly entered, but are normally omitted since one of the utility routines, such as ADDH, can be used to add them automatically.

If mistakes are made, they are easily corrected. For example, atom types, formal charges, and configurations can simply be relabelled correctly. Bond types can be changed by redrawing the bond and indicating the correct type. Atoms and/or bonds may be deleted by pointing at them while in the DELETE mode. The MOVE mode, on the other hand, allows for the movement of individual atoms.

Five commands are also available to manipulate the drawing on screen. These are: (1) TRANSLATE; (2) SHRINK; (3) EXPAND; (4) AUTOSCALE; (5) ROTATE. Various combinations of these commands will permit the user to adjust the graphical image in such a way that it remains on the screen.

MOLY also provides a number of utility routines for molecular storage, retrieval, orientation, and display. The following is a short summary of some of these utilities:

STORE: Stores a molecule on a file for use at a later date.
GET: Retrieves a previously stored molecule from a file.
DELM: Deletes a molecule from a file.
ORNT: Orients the molecule in space according to user-given specifications.
MROT: Rotates the molecule about the *X*, *Y*, and/or *Z* axes.
BROT: Alters the conformation of a molecule by performing rotations about acyclic bonds.
SETC: Sets the molecule into a conformation specified in terms of torsional angles.
REFG: Allows the user to modify the geometry by specifying new bond lengths, bond angles, and torsional angles.
INFO: Answers user questions about interatomic distances, angles, and torsional angles.
ADDH: Adds hydrogen atoms to the molecule in the proper positions.
DELH: Deletes all hydrogen atoms from the molecule.
MIRR: Reflects the molecule through the *XY* plane to create its mirror image (enantiomer).
INVE: Inverts specified stereocentres to create different stereoisomers.
SWAP: Exchanges the 'active' and 'inactive' molecules.
CPU: Reports the amount of computer CPU time used.
DPGM: Displays the bond lengths and angles.
DPCT: Displays the connection table.

We hope that the previous three examples have given you a flavour of molecular graphics systems and an appreciation of the problems involved. The range of available software is enormous: from a few lines of Fortran code to draw a simple stick figure to 20000 lines of code for the large graphics and analysis systems, such as MOLY.

10.5 SOME EXISTING SYSTEMS

We include the references to four systems that we believe to be generally available.

PLUTO, Crystallographic Data Centre, University Chemical Laboratory, Lensfield Road, Cambridge, UK, CB2 1EW

CHEMGRAF, Chemical Crystallography Laboratories, Oxford University (E.K. Davies and C.K. Prout).
MS, Quantum Chemical Program Exchange, Indiana University, Department of Chemistry, Bloomington, Indiana 47401, USA (The Molecular Surface program by Michael Connolly)
ORTEP 11, Oak Ridge Laboratory, Oak Ridge, Tennessee, Report ORNL-3704, USA (C.K. Johnson).

10.6 REFERENCES AND BIBLIOGRAPHY

Because of the nature of molecular graphics and the diverse fields of enquiry that are involved, such as computer science and computational geometry, we present an extensive, but by no means exhaustive bibliography of relevant papers. In selecting these various references we have attempted to cover a variety of systems, techniques and problems.

Anderson, W.F. and Cygler, M. 'Computer Modelling Studies of the Structure of a Repressor', *Bio Systems,* 18 (1) (1985), 3–14

Barry, C.D. (1973) 'Computer Graphics in Medicine and Biology', *Fourth International Biophysics Congress Proceedings Academy of Sciences of the USSR,* Moscow (1973), 366

— and North, A.C.T. 'The Use of Computer Controlled Display in the Study of Molecular Conformations', *Cold Spring Harbor Symposium on Quantitative Biology,* 36 (1971), 575

— *et al.* 'CHEMAST: A Computer Program for Modelling Molecular Structures', in *Information processing 71: Proceedings 1971 IFIP Conference,* vol 2, p. 1552 (North-Holland, Amsterdam, 1972)

— and McAlister, J.P. 'High Performance Molecular Graphics, A Hardware Review', in D. Sayre (ed.) *Computational Crystallography,* p. 274 (Oxford University Press, Oxford, 1982)

Blaney, J.M. *et al.* 'Computer Graphics in Drug Design: Molecular Modelling of Thyroid Hormone–Prealbumin Interactions', *Journal of Medical Chemistry* 25 (1982), 785

Blinn, J.F. 'A Generalization of Algebraic Surface Drawing', *ACM Transactions in Graphics,* 1 (3) (1982), 235

Blundell, T.L. 'Protein Crystallography, Interactive Graphics and Drug Design', SERC Bulletin, 2 (6) (1982), 8

Bond, P.J., (1972) 'Computer Graphics and Macromolecular Structures', *Computer Graphics ACM SIGGRAPH* 6 (1972), 13

Brandenburg, N.P. 'An Interactive Graphics System for Comparing and Model Building of Macromolecules', *Journal of Applied Crystallography,* 14 (1981), 274

Britton, E.G. 'Making Nested Rotations Convenient for the User', *Computer Graphics,* 12 (3) (1978), 222.

Cantor, C.R. and Schimmel, P.R. *Biophysical Chemistry, part 1, The Conformation of Biological Macromolecules* (W.H. Freeman, San Francisco, 1980)

Chabot, A.A. *et al.* 'Examples of the use of Interactive Computer Graphics with Real-time Energy Calculations in Studies of Protein–Ligand Interactions and Drug Design', in *Twelfth International Congress of Crystallography,* Abstract 18 (1981), 5–01

— *et al.* 'Examples of the Use of Interactive Computer Graphics with Real-time Energy Calculations in Modelling Molecular Interactions', in Griffin and Duax (eds) *Molecular Structure and Biological Activity* p. 399 (Elsevier Biomedical, Amsterdam, 1982)

Chothia, C. and Lesk, A.M. 'Helix Movements in Proteins', *Trends in Biochemical Sciences,* 10 (3) (1985)

Clark, D.D. and Schuster, S.M. 'Microcomputer Manipulation and Graphic Display of

Molecular Structures 1: Introduction; 2: File Structure and Maintenance; 3: Display', *Computers and Chemistry,* 4 (1980), 75, 79, 83
Cole, G.M. *et al.* 'Modelling Receptor and Substrate Interactions', in Olsen and Christofferson (eds), *Computer-assisted Drug Design, American Chemical Society,* Symposium 112, (1979), 190
Connolly, M.L. 'Solvent Accessible Surfaces of Proteins and Nucleic Acids', *Science,* 221 (1983), 709–13
Corey, E.J. and Wipke, W.T. 'Computer-assisted Design of Complex Organic syntheses', *Science,* 166 (1969), 178
Diamond, R. 'Some Problems in Macromolecular Map Interpretation', in R. Diamond *et al.* (eds) *Computing in Crystallography* (Indian Academy of Sciences, Delhi, 1981), 20 01.
— 'Inter-active Graphics', in R. Diamond *et al,* (eds) *Computing in Crystallography* (Indian Academy of Sciences, Delhi, 1981), 27 01.
— 'A Technique for Overlaying Common Storage', in R. Diamond *et al.* (eds) *Computing in Crystallography* (Indian Academy of Sciences, Delhi, 1981) 29 01
— 'BILDER: An Interactive Graphics Program for Biopolymers', in *Twelfth International Congress of Crystallography,* Abstract 18 5–02 (1981)
— 'Two Contouring Algorithms', in D. Sayre (ed.) *Computational Crystallography* p. 266. (Oxford University Press, Oxford, 1982)
— *et al* 'Three-dimensional Perception for One-eyed Guys', in D. Sayre (ed.) *Computational Crystallography,* p. 286 (Oxford University Press, Oxford, 1982)
— 'BILDER: An Interactive Graphics Program for Biopolymers', in D. Sayre (ed.) *Computational Crystallography,* p. 318. (Oxford University Press, Oxford, 1982)
Dubois, J.E. 'DARC System in Chemistry', in W.T. Wipke *et al.* (eds) *Computer Representation and Manipulation of Chemical Information,* p. 239 (John Wiley and Sons, Chichester, 1974)
Dyott, T.M. *et al.* 'MOLY: An Interactive System for Molecular Analysis'. *Journal of Chemical Information and Computer Science,* 20 (1980), 28.
Feldman, R.J. *et al.* 'An Interactive, Versatile. Three-dimensional Display, Manipulation and Plotting System for Biomedical Research'. *Journal of Chemical Documentation,* 12 (1974) 234
Feldman, R.J. 'The Design of Computine Syustems for Molecular Modeling', *Annual Review of Biophysical and Bioengineering.* 5 (1976), 477.
—, *et al.* 'Interactive Computer Surface Graphics Approach to Study of the Active Site of Bovine Trypsin', *Proceedings of the National Academy of Sciences, USA,* 75 (1978), 5409.
Ferrin, T.E. and Langridge, R., (1980) 'Interactive Computer Graphics with the UNIX Time-sharing system'. *Computer Graphics,* 13 (1980), 320.
Fox, J.L. 'Computer Graphics Aid Study of Molecules', *Chemistry and Engineering News* (21 July 1980), 27.
Glick, M.D. *et al.* 'Interactive Graphics for Structural Chemistry', *Computers and Chemistry* 1 (1976), 75.
Greer, J. and Bush, B.L. 'Macromolecular Shape and Surface Maps by Solvent Exclusion', *Proceedings of the National Academy of Sciences, USA,* 75 (1978), 303.
Gund, P. *et al.* 'Three-dimensional Molecular Modeling and Drug Design'. *Science.* 208 (1980), 1425.
Hardman, K. (1980) 'Database Requirements for Graphical Applications in Biochemistry', in Blaser (ed.) *Database Techniques for Pictorial Applications* (Springer-Verlag, Berlin, 1980)
Hermans, J. and Ferro, D. 'Representation of a Protein Molecule as a Tree and Application to Modular Computer Programs which Calculate and Modify Atomic Coordinates'. *Biopolymers,* 10 (1971), 1121.
Hodges, D. *et al.* 'MOLOCH-3: The G.D. Searle Molecular Modeling System', *169th National Meeting of the American Chemical Society,* Abstract COMP-7 (1975)
Humblet, C. and Marshall, G.R. Three-dimensional Modelling as an Aid to Drug Design', *Drug Development and Research,* 1 (1981), 409
Jacobi, T.H. *et al.* 'Molecular Modelling System', *Journal of Molecular Biology,* 72 (1972), 589.

Jones, T.A. 'A Graphics Model Building and Refinement System for Macromolecules', *Journal of Applied Crystallography*, 11 (1978), 268.
—, 'FRODO: A Graphics Fitting Program for Macromolecules', in D. Sayre (ed.) *Computational Crystallography*, p. 303 (Oxford University Press, Oxford, 1982)
Kan, L.S. *et al*, 'Computer Programs for Nucleic Acid Studies: Atomic Coordinates of Helices and their Graphic Display', *Computer Programs in Biomedicine*, 10 (1979), 16
Katz, L. and Levinthal, C. 'Interactive Computer Graphics and Representation of Complex Biological Structures', *Annual Review of Biophysics and Bioengineering*, 1 (1972), 465
Knowlton, K. (1981) 'Computer-aided Definition, Manipulation and Depiction of Objects Composed of Spheres', *Computer Graphics*, 15 (4) (1981), 352
— and Cherry, L. 'ATOMS, A 3-D Opaque Molecule System for Color Pictures of Space-filling or Ball-and-stick Models', *Computers and Chemistry*, 1 (3) (1977), 161
Kuntz, I.D. *et al.* 'A Geometric Approach to Macromolecule–Ligand Interactions', *Journal of Molecular Biology* 161 (2) (1982), 269
Langridge, R. 'Interactive Three-dimensional Computer Graphics in Molecular Biology', in *Ninth International Congress of Crystallography*, Abstract XXV-8 (1972)
— and MacEwen, A.W. 'The Refinement of Nucleic Acid Structures', in IBM *Scientific Computing Symposium on Computer-aided Experimentation* (IBM, Yorktown Heights, New York, 1965), 395
— *et al.* 'Real-time Color Graphics in Studies of Molecular Interactions', *Science*, 211 (1981), 661
Lesk, A.M. 'Themes and Contrasts in Protein Structures', *Trends in Biochemical Sciences*, 9 (6) (1984)
— and Hardman, K.D. 'Applications of Large-scale Computers and Computer Graphics: Investigation of Biological Macromolecular Structure, Function and Evolution', in *Supercomputers in Chemistry*, American Chemical Society, Symposium 173 (1981), 143
— and Hardman, K.D. 'Computer-generated Schematic Diagrams of Protein Structures', *Science*, 216 (1982), 539
Levinthal, C. 'Molecular Model-building by Computer', *Scientific American*, 214 (1966), 42
Levitt, M. and Feldman, R.J. 'Conformational Dynamics of Pancreatic Trypsin Inhibitor: A Movie', in M. Balaban (ed.), *Structural Aspects of Recognition and Assembly in Biological Macromolecules*, (Balaban International Science Services, Rehovot, 1981), 35
Marshall, G. and Marshall, R. 'CHEMAST and Macro-molecular Modeling System: Two Different Approaches to Molecular Modeling', *Ninth International Congress of Crystallography* (1972) Abstract XXV-9
— *et al.* 'Macromolecular Modeling System: the Insulin Dimer', *Diabetes*, 21, Suppl 2 (1972), 506
— *et al.* 'Computer Modeling of Chemical Structures: Applications in Crystallography, Conformational Analysis and Drug Design', in W.T. Wipke *et al.* (eds) *Computer Representation and Manipulation of Chemical Information*, p. 203 (John Wiley and Sons, Chichester, 1974)
Max, N.L. 'ATOMLLL: Atoms with Shading and Highlights', *Computer Graphics*, 13 (2) (1981), 165
— *et al.* 'Computer Graphics and the Generation of DNA Conformations for Intercalation Studies', *Computers and Chemistry* 5 (1) (1981), 19
McLachlan, A.D. 'Rapid Comparison of Protein Structures', *Acta Crystallography A.* (A38) (6) (1982), 871.
Meyer, E.F. 'Three-dimensional Graphical Models of Molecules and a Time-slicing Computer', *Journal of Applied Crystallography*, 3 (1970), 392
— 'Interactive Computer Display for the Three-dimensional Study of Macromolecular Structures', *Nature*, 232 (1971), 25
— 'Storage and Retrieval of Macromolecular Structural Data, *Biopolymers*, 13 (1974), 419
— 'Interactive Graphics in Medicinal Chemistry', in Ariens (ed.) *Drug Design (IX)* (Academic Press, New York 1980), 267
— and Willoughby, T.V. 'Three-dimensional Graphics with a Minicomputer and a Television Raster Display', in *Ninth International Congress of Crystallography* (1972), Abstract XXV-6
— *et al* 'CRYSNET, A Crystallographic Computing Network with Interactive Graphic

Display', *Federation Proceedings*, 33 (1974), 2402
Morffew, A.J. and Moss, D.S. 'A Strategy for Combining Restrained Least Squares with Computer Graphics in the Refinement of Protein Structures', *Acta Crystallography A.* (A39) (1983), 196
— *et al*, 'The Use of a Relational Database for Holding Molecule Data in Molecular Graphics System', *Computers and Chemistry*, 7 (1) (1983), 9
Musso, J.A. 'A Three-dimensional Perspective Plotting Routine (Application to Molecules)', *Computers and Chemistry*, 4 (1980), 149
Nagano, K. 'How to Simulate Protein folding with Interactive Computer Graphics', in Ahmed (ed.) *Chrystallographic Computing Techniques* p. 344 (Munksgaard, Copenhagen, 1976)
Nichol, C.J. *et al.* 'Half-tone Representation of Molecular Surfaces', *IUCC Bulletin*, 4 (1982), 1
Nix, C.L. and Rubin, B. 'Interactive Molecular Dynamic Display', *Journal of Applied Crystallography*, 15 (1982), 467
North, A.C.T. and Barry, C.D. 'Molecular Modelling and Computer Graphics', *Biochemical Journal*, 121 (1971), 121.
— *et al.* 'The Use of an Interactive Graphics System in the Study of Protein Conformations', in Srinivasan (ed.) *Biomolecular Structure, Conformation, Function and Evolution*, p. 59 (Pergamon, Oxford, 1981)
Olson, A.J. 'GRAMPS: A High Level Graphics Interpreter for Expanding Graphics', in D. Sayre (ed.) *Computational Crystallography*, p. 326 (Oxford University Press, Oxford, 1982)
Ortony, A.A. 'A System for Stereo Viewing', *Computer Journal*, 14 (1970), 140
Porter, T. 'The Shaded Surface Display of Large Molecules', *Computer Graphics*, 13 (2) (1979), 234
Richardson, J.S. 'The Anatomy and Taxonomy of Protein Structure', *Advances in Protein Chemistry*, 34 (1981), 167
Richards, W.G. and Sackwild, V. 'Computer Graphics in Drug Research', *Chemistry in Britain*, 18 (9) (1982), 635
Smith, G.M. and Gund, P. 'Computer-generated Space-filling Molecular Models', *Journal of Chemical Information and Computer Science*, 18 (1978), 207
Staudhammer, J. 'On Display of Space-filling Atomic Models in Real-time', *Computer Graphics*, 12 (3) (1979) 167.
Takenaka, A. and Sasada, Y. 'Computer Manipulation of Crystal and Molecular Models', *Journal of the Chrystallography Society of Japan.* 22 (1980), 214
Tsernoglou, D. *et al* 'Protein Sequencing by Computer Graphics', *Biochemical and Biophysical Acta*, 491 (1977), 605
Willoughby, T.V. and Meyer, E.F. 'Fourier, Contouring and Molecular Modelling Using Three-dimensional Computer Graphics', in *Ninth International Congress of Crystallography* (1972) Abstract XXV-7.
Zientara, G.P. and Nagy, J.A. (1983) 'Proteins and Polypeptides: Computer Graphics for Space-filling Model Representations', *Computers and Chemistry*, 7 (1983) 6.

11 Simulation and Animation

11.1 MOVING PICTURES

If you have read through the earlier chapters of this book you will have already learnt many of the fundamental techniques of graphics programming. Some of these techniques involve the most straightforward kinds of movement of graphic data: transformations like rotation, scaling, and translation. You have seen for example in Chapter 10 that application of rotations in real-time can give a dynamic picture of the structure of molecules. The present chapter will consider the general question of how movement of graphic images can be carried out in real-time. Examples of both research simulations and animations suitable for teaching purposes will be discussed with reference to a number of different graphics systems.

Biologists use simulations to study the workings of complex dynamic processes. By abstracting elements from the real system and producing a computer model, the behaviour of the model can be compared with data from the real system. In this way, the success of the model may be gauged, and adjustments to model parameters can be made as required. The assumptions made during model construction can then be used as a starting point for further investigation of the biological system itself. In short, simulation modelling gives insight into the workings of real systems. Many examples of simple simulations in the life sciences are to be found in the book by Spain (1982).

Early simulations tended to produce numerical output only, and as such work was largely batch-oriented, the degree of interaction between programmer and program was very limited. The program development period was very slow, as the numerical output had to be analysed at each stage. With the rise in popularity of interactive, online computing, graphic methods became essential; it is a waste of time to have to do an hour's data analysis before amending a program under development. Use of graphical output in the form of graphs, histograms or pictures often enable the programmer to see 'at a glance' how the model is behaving. Alterations or changes in data can then be made almost instantaneously.

Let us first of all dispense with the graphically most simple class of simulation: simulations where a single static set of results are obtained. Although data produced in simulations of this type can involve production of complex graphs, the graphic problems involved are those of display of a single picture, and we have already looked at techniques for this kind of data analysis in the earlier chapters.

Often, however, the processes to be studied produce changes over time, and so we need to use graphic techniques for temporal modification of the outputted images. Our main concern here will be with CRT output; although 'time slice' output of individual images from simulations can of course be outputted to a plotter or graphics printer, the problem again reduces to output of static images.

11.2 HARDWARE FOR REAL-TIME ANIMATIONS

The phrase 'real-time animation' refers to animations that are performed at the same time as the program that generates them is run. This is in contrast to animations that consist of a 'replay' of stored graphic data, and the differences between these two categories will be considered below.

Graphic animation involves the erasure of existing picture elements and the display of new ones (Catmull, 1978; Walton and Risen, 1969; Wein and Burtnyk, 1976). The first limitation that must be taken into account is whether or not the graphics device to be used is capable of selective erasure. The majority of graphics CRT displays available to the biologist will probably fall into this category, with the exception of storage tube (or storage tube emulating) devices: these displays were described in Chapter 2.

The alternative to selective erasure is erasure of the whole screen (using a 'newpage' command; see Chapter 3). This compromise is sometimes acceptable in animation work, if all or most of the image is to be redrawn and the display can draw lines very quickly. Problems occur if only a small part of the picture needs to be updated and the drawing speed of the device is slow. The second limitation to real-time work is the speed of drawing each frame of the animation. Almost any computer with a non-storage tube graphics CRT display can draw a line between two endpoints, erase it and redraw it between two new endpoints at 'almost instantaneous' speed. The problems begin if a picture composed of hundreds of lines is to be drawn; in this case, the capabilities of the hardware become of prime importance. As you saw in Chapter 2, random scan CRTs are usually capable of drawing a very high resolution image with a large number of lines more quickly than an equivalent refresh display. High quality random scan displays are not cheap, and it is often better (at least from the financial point of view) to settle for a lower resolution refresh display to do the same task.

The display is not the only factor limiting speed of image creation. Inefficient software and computer architecture can also cause problems. A small microcomputer using interpreted Basic as the programming language is not capable of the processing speed offered by a compiled language, or code written on the same microcomputer in assembly language. The word length of the computer will also be important, as transfer of data along an 8 bit data bus will be, in general, less 'efficient' than along a 16, 32 or even

64 bit bus. It is important, therefore, that the capabilities of a particular graphics system are well understood before embarking on anything but a trivial animation sequence. It is a waste of time to spend several weeks attempting to coax impossible performance out of an inferior graphics system.

11.3 CONCEPTS OF GRAPHIC ANIMATION

Animation techniques consist of drawing and redrawing whole or part images to give an impression of movement. If the movement is some kind of regular transformation — a rotation for example — the positions of all graphics primitives on the screen are recalculated for each increment to be displayed. Simulations on the other hand may involve appearance or disappearance of picture elements as the animation proceeds. Although 'real-time' animations are often the most useful form of display for simulations, they often need a powerful computer to perform the computations from which the graphic data is obtained. It is not encouraging or constructive to wait for eight hours in front of a CRT to see the next frame of the simulation! In such cases the experimenter would probably store the graphic images in a set of data files to be 'played back' as a more or less continuous series of frames. This technique relies on use of a graphics CRT with extremely rapid line-drawing capability, but an alternative method is to use a movie camera system to photograph the course of the simulation. Software-controlled camera systems are available for a variety of graphics systems. This technique has been used to good effect with the cell division simulations described later in this chapter.

Drawing and Redrawing the Picture

In most cases the impression of movement is given by selective erasure and redrawing of picture elements. A simple flow diagram for a single animation step could be as follows:

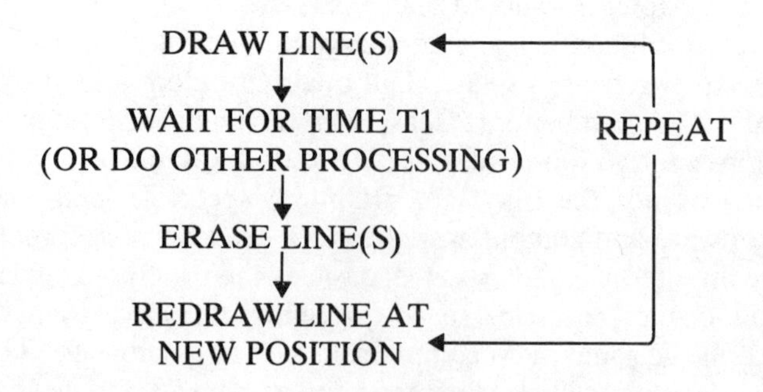

The success of this procedure depends heavily on the speed with which the line can be erased and redrawn. However, once a picture segment has been displayed at a given set of coordinates on the screen it should not be moved too quickly, or the eye will not be able to register the information before it is erased. These 'timing calculations' are clearly affected by the particular graphics hardware and software in use. In some cases the speed of drawing and the time taken for the computation of the graphics data may mesh together and obviate the need for timing calculations. This is the exception rather than the rule.

You may wish to program the following example in order to try out the effect of timing and to check the speed of line drawing using the graphics device available to you. Write a short program to perform the following operations.

(1) Calculate two (x,y) pairs at random within the screen coordinate limits of the display in use.
(2) Draw a line between these pairs.
(3) Pause (as necessary).
(4) Erase the line.
(5) Loop back to step 1.

The goal of this example is to produce smooth movement that fools the eye into believing that the line has shifted itself, and is translating and rotating in three-dimensional space. A more testing gauge of your graphics programming skills would be to take a line of fixed length and to rotate it by perhaps 5 degrees for each step of the program.

In the rest of the present chapter we will look in detail at three different simulation/animation 'case studies'. The first will be a simple 'dynamic graph' simulation, followed by simulation of cell division processes during animal development. The final example is a 'teaching' animation of the processes involved in transcription and translation in cells.

11.4 DYNAMIC GRAPH CONSTRUCTION

Perhaps the simplest animation of interest to the biologist is the display of two-dimensional graph data that is updated in real-time. You may recall that we looked at the construction of simple graphs in Chapters 1 and 7, and an animated version of such a graph is really very simple on a display capable of selective line erasure. An animated graph consists of a constant set of graphics primitives (axes, scale marks and labels), and a changing set of primitives (the markers for each data point and the lines drawn between the markers). At each time increment the following operations must be performed:

(1) Calculate new (x,y) coordinates for each point from the simulation equations.
(2) Enter the new (x,y) coordinates into a data structure available to the graphics processor.
(3) Erase the old primitives from the screen.
(4) Plot the new primitives using the newly calculated data.
(5) Loop back to step 1.

This process is identical to the simple line-moving program outlined above. Observation of a dynamically changing pattern on a graph can give useful insight into the process being modelled. If the amount of information in each frame is not large, it may be informative to omit the erasure phase, plotting all sequential plots on top of each other. This kind of dynamic graph display is identical in graphics terms to the display of data obtained over a course of time from a piece of laboratory equipment. In the latter case, a severe timing problem emerges with biological data: events either occur over milliseconds (ms) (for example, nerve impulses) or over minutes or hours. In both cases, techniques of data capture, and later replay using the graphic display, will be necessary.

Three-dimensional graphs make more use of the abilities of graphics devices to display information in a readily assimilated form, although the increased complexity of the information presented on screen make it more difficult to use animation techniques. Indeed, it is possible to use a three-dimensional plot to display changes of two variables over time, if time is itself plotted as the third axis. Various algorithms have been published enabling display of X,Y,Z data, and a simple algorithm was given in Chapter 7. Three-dimensional plots have been used to great effect in simulation work. A good example of a study of this kind is the work of Hans Meinhardt (1976; 1983) shown in Figure 11.1. Meinhardt studied the interaction of two substances (an activator and an inhibitor) hypothesised as determining the differentiation of cells in developing tissues. In order to study the behaviour of the model system in two dimensions, a three-dimensional graph system was used to plot activator concentration over a two-dimensional sheet.

A similar display technique has been used in our own laboratory to study strain fields in the developing tip of the single-celled marine alga *Acetabularia* (Goodwin and Briere, 1985). A modified finite-element method is used to describe the interactive dynamics of the cytoskeleton and calcium in the cortical cytoplasm. The *Acetabularia* tip may be mathematically described as a three-dimensional shell, and for display purposes, the shell is 'unwrapped' into a rectangular mesh made up of a number of finite elements (Figure 11.2). Figure 11.3 shows an example simulation result in which the variation of calcium concentration is shown over the shell.

Figure 11.1: Simulation of the Generation of a Branching Filament (after Meinhardt, 1976). The *X* and *Z* axes represent a two-dimensional space, and the *Y* axis is concentration of an activator substance. A new activator peak is formed at a critical distance behind the growing point (a,b) initiating a lateral branch (c) which elongates. When the tip of this branch has attained a sufficient distance from the original filament (d) a new branch can be formed

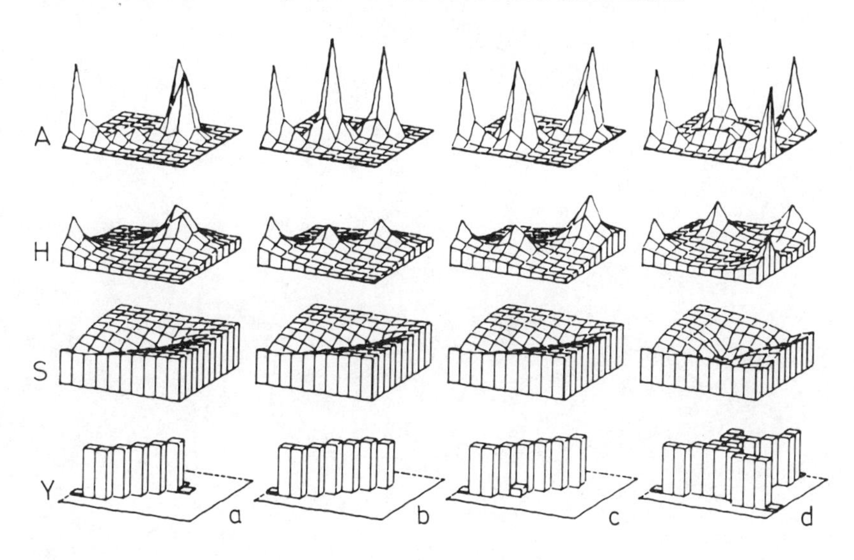

11.5 SIMULATION OF CELL DIVISION AND CELL INTERACTION PROCESSES

Alphanumeric Character-based Simulations

Cell division provides a useful model system for analysis of how computer graphics are used in simulation work. The reason for this is partly that the patterns taken up by dividing cell masses consist of shapes in space, and so provide a physical rather than abstract graphic modelling task. Cell division systems have also been studied using a variety of graphic methods, and a review of these methods also gives valuable information about graphics simulations.

The use of simulation techniques to model the spatial aspects of cell division processes have their origins in the cellular automata of von Neumann (1953) and the subsequent automata of Maruyama (1963) and Ulam (1962). These model systems consider cells as occupied 'squares' on a two-dimensional 'chessboard'. Cell division is simulated by filling a square adjacent to the dividing cell with a daughter cell (Figure 11.4). By setting conditional rules for the division process, different cell patterns were obtained. These abstract cell models were later refined by various authors including Ede and Law (1969), who used a model system of this type to simulate limb bud growth, Ransom (1975, 1977), who modelled

Figure 11.2: Unwrapping of the Ellipsoid Used to Simulate Pattern Formation in the Growing Tip of Acetabularia. Top — the three-dimensional structure; centre — unwrapping of the ellipsoid along the line *a b* to *c d*; bottom — the two-dimensional sheet obtained

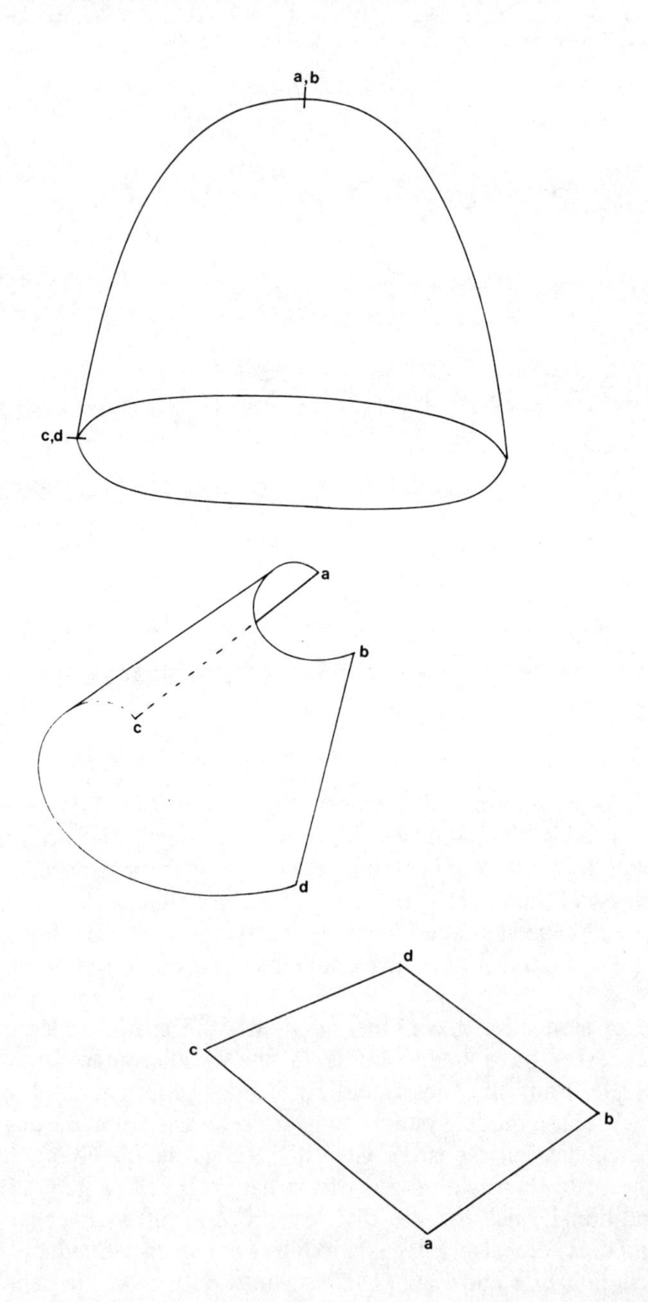

Figure 11.3: Simulation of Calcium Concentration in the Growing Tip of *Acetabularia*. The *X* and *Z* axes represent the two-dimensional sheet, and calcium concentration is indicated in the *Y* axis

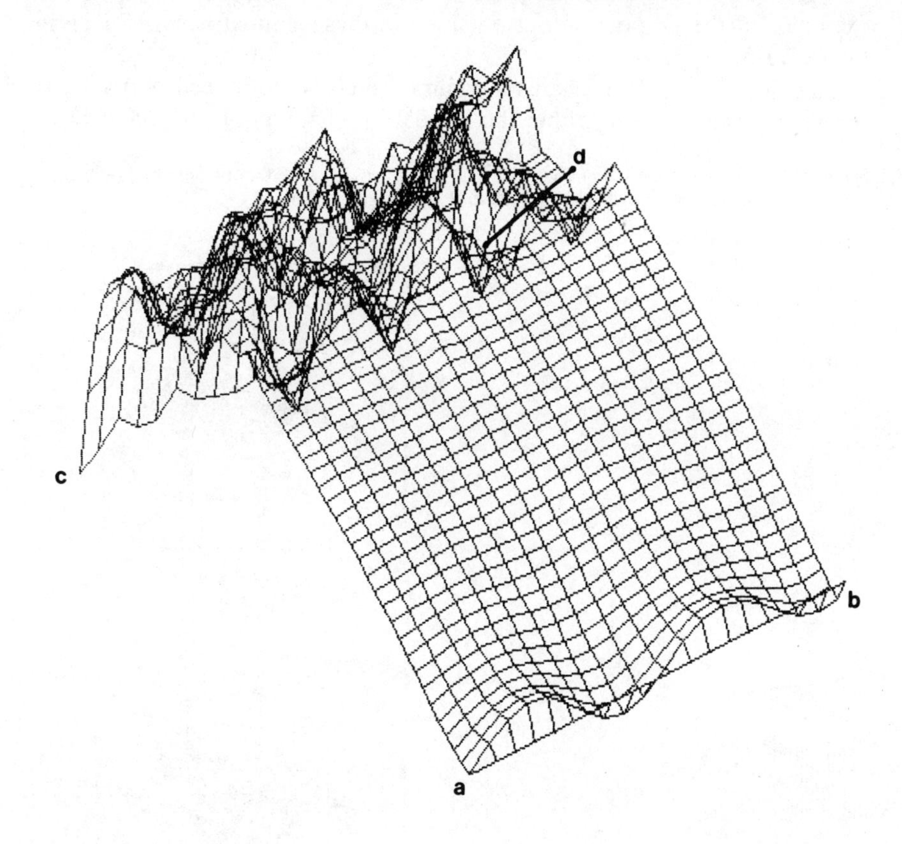

Figure 11.4: Simulation of Cell Division on a Square Tessellated Grid. The single cell (left) 'divides' by occupying a neighbouring square (centre). Successive divisions produce a cluster of 'cells' (right)

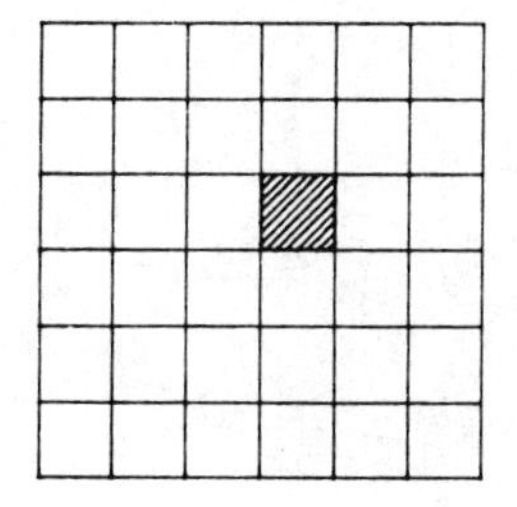
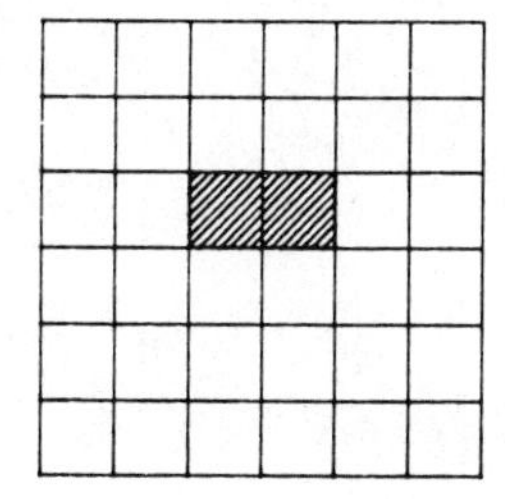
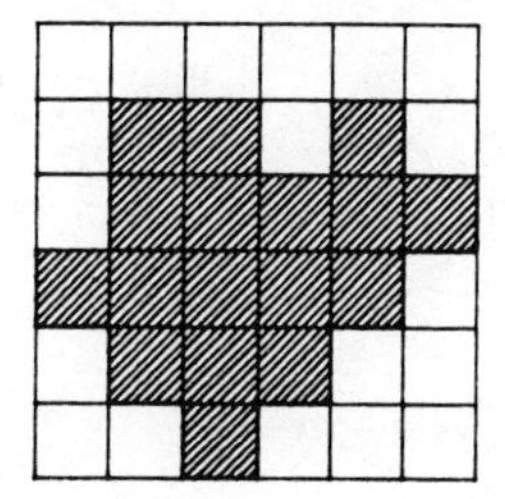

insect imaginal disc development, and Düchting and Vogelsaenger (1981, 1983) who modelled tumour growth in three dimensions. Some examples of the graphic output obtained in these simulations are given in Figure 11.5.

This group of models shares a common methodology; cell populations are all represented using either the standard alphanumeric character set of

Figure 11.5: Cell Simulation Using Character Representation. (a) A vertebrate limb bud (Ede and Law, 1969); (b) tumour growth (Düchting and Vogelsaenger, 1981)

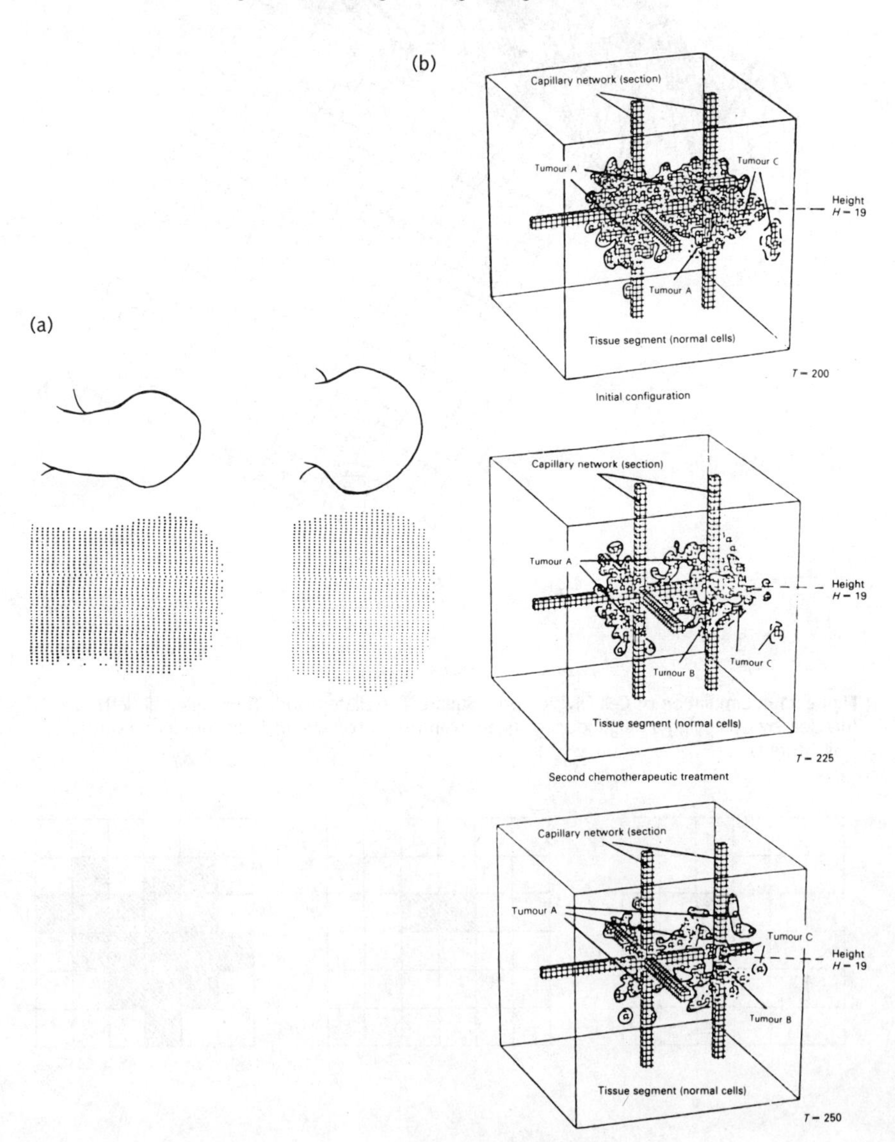

a printer or CRT, or perhaps special symbols, arranged in a regular square or hexagonal tessellation. If we take the tessellation to be a representation of the rows and columns of alphanumeric output, then the simulation becomes graphically extremely simple to program. The command 'divide cell to the left' is encoded as 'occupy tessellation position at the left of the position occupied by the cell to divide', and a two-dimensional integer or byte array becomes a logical, if rather inelegant data structure to hold the tessellation (Figure 11.6). Movement of cells around the tessellation can be modelled by transfer of the numeric values representing type and identity of a cell from location to location.

To our knowledge, the insect cell division model developed by Ransom (1975, 1977) was the only version of the early cell division simulations to use a graphics CRT for real-time analysis. (The technique of using characters to represent abstract cells or single-celled organisms predated this work, as the well-known 'Game of Life', a less-serious version of the Ulam automata, was animated many times in the early 1970s). Use of a refresh CRT for this type of application involves entering the character data to be displayed into the segment of display memory holding the character screen map (screen mapping is discussed in Chapter 2). Although this process sounds simple, it can be quite a complex matter to control alphanumeric output in this way. It is often an easier proposition on a cheap home computer than on an expensive mainframe computer with a high quality CRT. This is because the display memory on a small microcomputer can be

Figure 11.6: Relationship Between a Two-dimensional Array and a Tessellation. The rows and columns of a two-dimensional array are shown indicated along the edges of the tessellation. Array locations marked '1' contain 'cells', those marked '0' are 'empty'

	1	2	3	4	5	6	7	8	9	10
1	0	0	0	0	0	0	0	0	0	0
2	0	0	0	0	0	0	0	0	0	0
3	0	0	0	0	0	0	1	0	0	0
4	0	0	0	0	1	1	1	1	0	0
5	0	0	0	1	1	1	1	0	0	0
6	0	0	0	0	1	1	0	0	0	0
7	0	0	0	0	0	1	0	0	0	0
8	0	0	0	0	0	0	0	0	0	0
9	0	0	0	0	0	0	0	0	0	0
10	0	0	0	0	0	0	0	0	0	0

accessed directly from an interpreted Basic program (or often using a compiled language). To perform the same task on a mainframe computer coupled to a refresh display would involve writing machine code routines to address the character screen locations on the display.

Line Representations

If cells are represented by a regular tessellation, character animation is the most efficient and easily programmed methodology. Biological cells are not regular shapes with four, six or eight sides, and more complex cell models have been developed which allow cell shape and neighbour number to be varied. Alphanumeric characters are not suitable for use in such cases, and different animation strategies have to be employed. In this subsection we will consider one approach to modelling cells with irregular shapes.

The topological model for cell sorting (Matela and Fletterick, 1979; 1980) proposed the use of graph theory to represent cells (Figure 11.7). In this system, each cell is represented by a node, with lines representing adjacencies between neighbouring cells. There is an equivalence between a graph of this type (which contains no inherent spatial information) and its geometric dual, a map which contains information on the geometry of the system (dotted lines in Figure 11.7). Note that the graph is triangulated.

Figure 11.7: Simulation of Cells Using a Topological Graph Representation. Each cell is represented by a 'node' shown here as a circle. Solid lines show connectivities ('cell adjacencies') between nodes. Dotted lines show schematic representation of cell boundaries

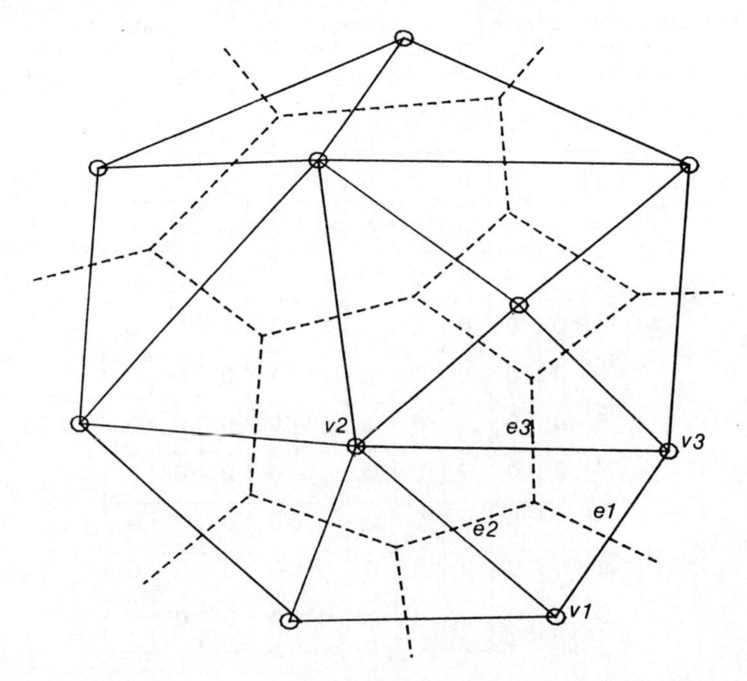

This form of representation ensures that three cells always meet at corners — the most common type of junction in real cell systems. One of the most powerful features of this graph model is its ability to simulate changes in cell contacts. This is facilitated by the exchange rule, shown in Figure 11.8, and the repetitive application of this mechanism allows the model to simulate cell sorting.

More recent versions of the graph model have also included simulation of cell division (Matela, Ransom and Bowles, 1983; Ransom and Matela, 1984; Matela and Ransom, 1985). Cell division is modelled by use of a series of division masks (Figure 11.9), masks being available for cells with from four to nine neighbours. If a cell is to divide, the appropriate mask is called and rotated as required. The data structures holding the nodes and adjacencies are then updated to include the new node and its connections to neighbouring cells.

The earliest simulations using the graph model did not employ graphic output at all, and all reconstructions were performed from numeric data. This led to frayed nerves and use of primitive graphic devices (Figure 11.10). The graph system is, however, ideally suited to computer graphic analysis, as only points and lines are to be drawn. The main difficulty lies in imbibing the graph model with geometrical information so that the dual can be constructed in two-dimensional space. As the graph does not possess coordinate data *per se*, a set of rules governing node positions have to be created.

If we accept the value of graphic output of the cell system, two options are opened to us. First, the (x,y) coordinate information can be stored and updated within the computer, and the data is displayed only at selected intervals during the course of the simulation. This is inferior to the display

Figure 11.8: The Exchange Rule for Cell Connectivities. The bond between cells *B*1 and *W*1 (a) may be exchanged for the bond between cells *W*2 and *W*3 (b)

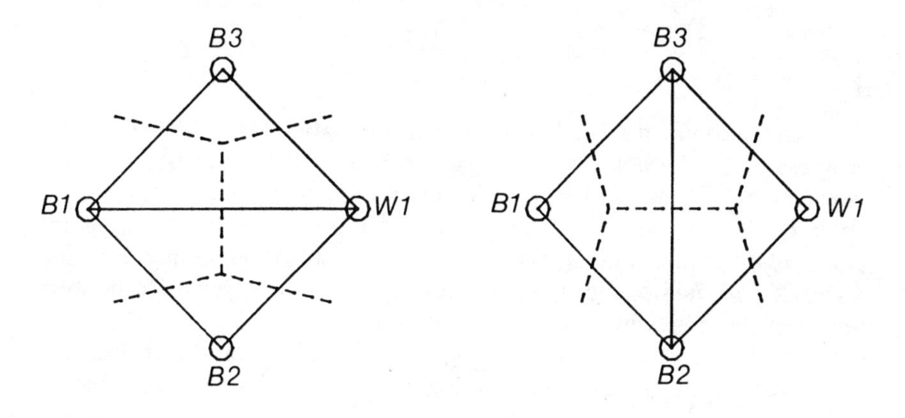

Figure 11.9: Division Patterns for Cells with Between Four and Nine Neighbours. In each case the upper diagram represents adjacencies on graph and dual before division. The lower diagram shows adjacencies after cell division has taken place

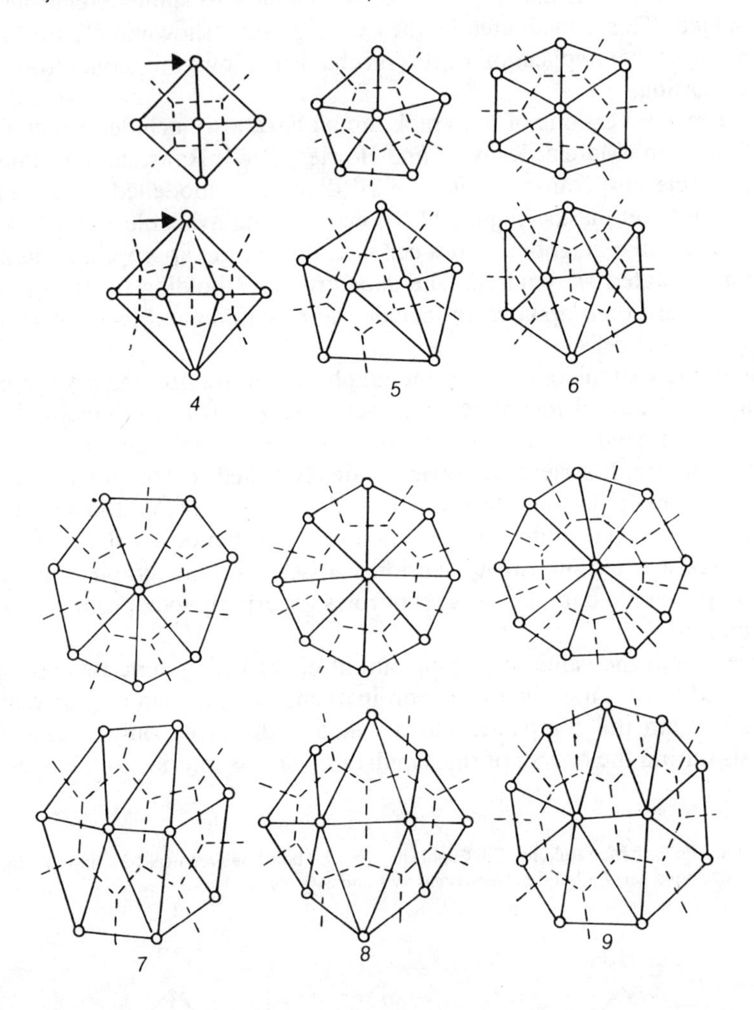

of an animation of the whole cell division simulation, as little 'feel' for the progress of the division dynamics is given. It is also easier to trap errors or shortcomings in the program if a continuous sequence of divisions is observed. There are several options for continuous display of the graph model in operation. The first is to display only the graph, in monochrome. As each cell division occurs, the lines representing adjacencies between cells and the positions of the nodes representing the cells are erased and redrawn. This method is easy to program, but has several drawbacks. The first of these is the difficulty in 'spotting' each division as it occurs. Changes

Figure 11.10: Hand Reconstruction of Graph Data for the Cell Division Model

in, say, ten lines in a total of several thousand will often go unnoticed unless some method of highlighting the dividing cell is used. In monochrome, the options are limited. If only one cell type is present in the system, the dividing cell could be temporarily displayed with dashed or dotted lines instead of solid lines. This may be impossible for more than one cell type, as dashed lines may already be used to indicate connections between cells of different type. An alternative would be to 'flash'; the lines to be updated by erasing and redrawing them for a few seconds.

It is also difficult to observe behaviour of mixed cell populations using a monochrome representation of the growing graph. As we have just seen, the use of dashed patterns between nodes of different types allows a crude boundary between cell types to be displayed, but a boundary of this type can be very confusing, for example if cells are present in small clumps surrounded by cells of a different type, or if more than two types of cell are present. A solution for growth of two cell types is to draw the nodes as circles and to block in the nodes representing cells of one type. Alternatively, two different markers could be used to represent the node positions.

All of these problems are overcome if a colour display is available. Nodes and connectivities between nodes can be drawn using a suitable colour code, and as more than three or four colours would rarely be needed, even a display with a limited range of colours can be used.

Figure 11.11: Graph Model of Cell Division; Simulation of the Sorting Between Two Growing Populations of Cells (From Ransom and Matela, 1984). In (a) no sorting has occurred along the boundary between the two cell populations; in (b) sorting of the two cell types has occurred using the exchange rule (see Figure 11.8)

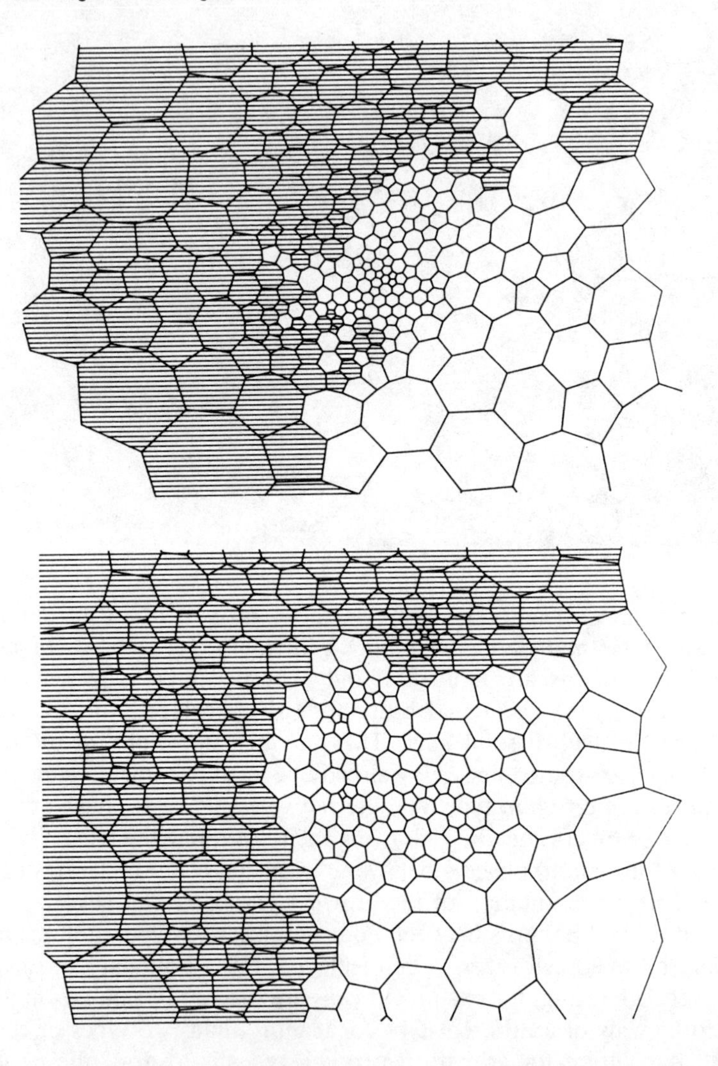

As the graph is basically a system giving information on the topology, a better display solution is to forget the graph altogether and draw only the dual. This gives a better approximation to the division of 'real' cells with each cell boundary represented. The clearest method for highlighting the different cells is to define the boundaries defining each cell as a polygon, and to then use a polygon filling routine to mark each cell, a different pattern or colour being used for each cell type. Using the Tektronix Plot 10

Interactive Graphics Library (IGL), the endpoints of the lines defining a polygon (cell) can be entered into two one-dimensional arrays which are then used as data to draw the polygon using the POLY command, thus:

CALL POLY (code)

called from within a Fortran program linked to the IGL library will draw a polygon of ten points using as data the first ten elements of the arrays X(i) and Y(i). Also using IGL, the polygon may be filled using the command:

CALL PANEL (code)

Some typical examples of output from a version of the program using this method of displaying cells are shown in Figure 11.11 and Plate 6.

There are two options for observing the simulation in progress. The first is to use a powerful computer and a suitable colour graphics display to observe the graphic output in real-time. As we saw earlier in this chapter, this option may be impossible to use because of hardware constraints. The alternative is to use a movie camera system attached to the graphic device. As each cell division is generated on the display it is photographed, and the course of the whole simulation may be observed after the film is processed. Some trial and error is needed with this procedure, as a number of movie frames of the same display picture will be required if the film is to run at a suitable simulation speed. We have used a Dunn camera system to make movies of the cell division simulation. This system includes a Bolex 16 mm movie camera, and we used News film exposing ten frames per image. This gave an acceptable speed for the simulation when the projector was run at 24 frames per second.

11.6 ANIMATION OF GENETIC EVENTS

In the previous section we saw that computer graphics can be used to illustrate the behaviour of a dynamic model. This application is very much a research tool, with the output providing valuable data on the effects of parameter variation. Graphic animations are not limited to the research area, however, and can also be used in teaching situations. Animation can provide the student with insight into the dynamics of biological processes, and it is likely that the use of computer graphics in this context will increase in the future. Here we will discuss animation in a representative program — animation of the process of RNA transcription and translation using a microcomputer (Ransom, 1986).

The first step in producing an animation is to decide on the elements to be abstracted from the process to be modelled. This step is identical to the model formulation step needed to devise a research simulation. The basic elements of our genetic simulations are the DNA and RNA bases, and we use the code letters A, C, G, T and U to describe these units. Another important class of elements are the amino acids to be joined into a polypeptide chain. These are denoted by the accepted three-letter abbreviations VAL, ARG, THR and so on. Other elements are enzyme and ribosome subunits, represented in the animation by coloured or shaded blocks.

The animation process has to move these elements relative to one another to produce an acceptable representation of the processes of transcription and translation. Once the elements of the animation have been decided on, a list of subprocesses can be made. In the case of transcription, for instance, such a list would include the following events:

(1) coupling of polymerase to DNA;
(2) assembly of the RNA chain on the DNA template;
(3) termination of RNA synthesis (including hairpin loop formation for prokaryotes).

In terms of animation, these subprocesses pose a number of graphics programming problems, namely:

(1) creation of blocks representing polymerase subunits;
(2) movement of these polymerase blocks;
(3) arrangement of DNA bases into chains;
(4) 'synthesis' of the RNA chain;
(5) movement of DNA and RNA chains relative to one another.

As the bases are to be represented as letters, these problems distill down to methods of moving text cells around on the screen. Consider Figure 11.12 showing a 'still frame' from an animation of transcription. If the nascent RNA chain moved from left to right along the DNA duplex it would quickly disappear off the right-hand edge of the screen. The alternative method is for the point of RNA synthesis to remain stationary while the DNA duplex moves to the left. Although this mechanism is distinctly 'unbiological', the net effect is to give the appearance of 'panning' from left to right. The newly formed RNA chain builds towards the right from the site of synthesis.

How is the movement physically carried out? A CRT screen can be divided into a number of text cell locations (commonly 40 × 25 or 80 × 25) as shown in Figure 11.13. In order to move a chain of bases from left to right across the screen, the contents of each text cell must be moved into the neighbouring right-hand cell. If the chain extends beyond the left-hand edge of the screen, new base values will need to be written into the left-most text cell for every right shift (Figure 11.14).

Figure 11.12: 'Still Frame' from an Animation of Transcription Using a Microcomputer; This Picture is a Screen Dump from a Commodore 64 Computer

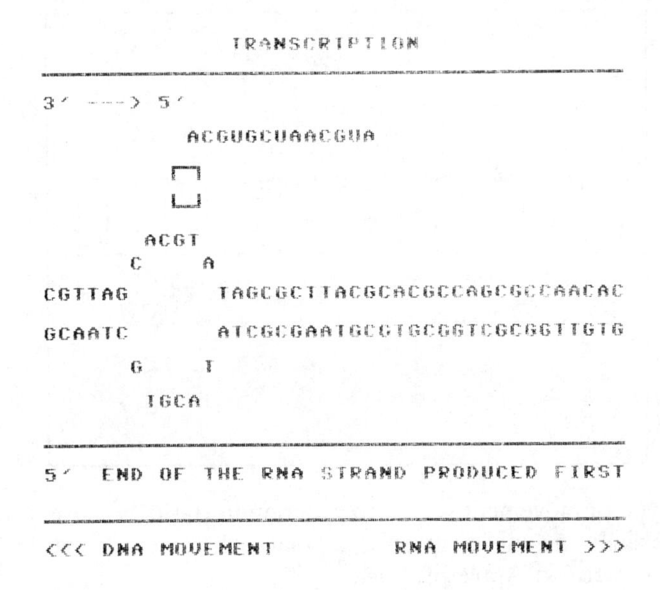

Figure 11.13: CRT Screen Showing Relationship of Display Area to Text Cells

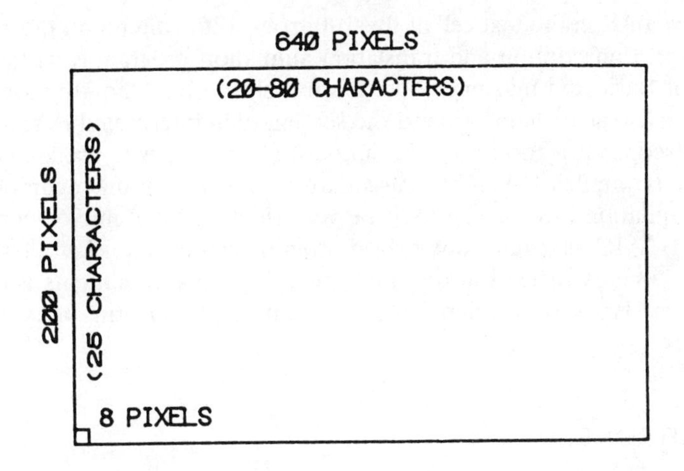

Figure 11.14: Movement of Chains of Characters Across the Screen. Each character occupies a text cell, and an algorithm to move the contents of each cell into the neighbouring right-hand cell is used

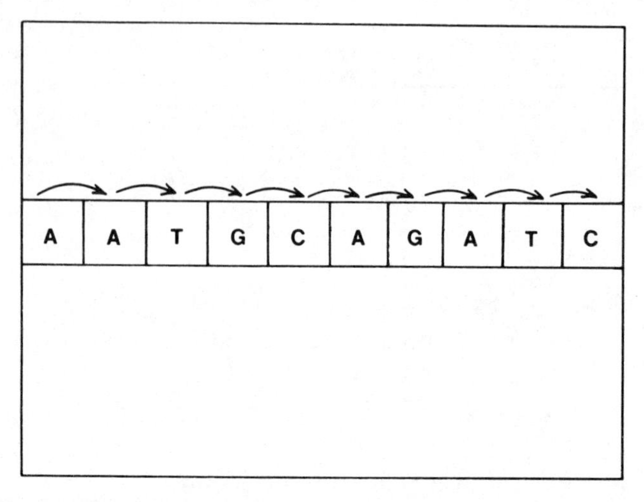

This type of movement is easy to program in Basic, because most interpreters usually include commands to place characters in particular text cells, for instance in Applesoft Basic:

```
10 VTAB(10)
20 HTAB(12)
30 PRINT"X"
```

will place an X in the text cell at the 10th row, 12th column on the screen.

The full transcription and translation animation program is written in a hybrid of Basic and machine code. If we wish to give 'smooth' movement of the nucleic acid chains around the screen, then interpreted BASIC cannot be used, as it is too slow. The simplest alternative is to use a compiled language (compiled BASIC or Pascal are most usual on microcomputers). The compilation process can itself be very slow on small microcomputers, and the BASIC/machine code hybrid program can be a quicker alternative in some circumstances. Luckily, movement of text cell contents is one of the easiest types of machine code operation to perform on a microcomputer.

11.7 REFERENCES

Catmull, E. 'The Problems of Computer-assisted Animation', *Computer Graphics*, 12 (3) (1978), 348–53

Düchting,W. and Vogelsaenger, Th. (1981) 'Three-dimensional Pattern Generation Applied to Spheroidal Tumor Growth in a Nutrient Medium, *International Journal of Bio-Medical Computing*, 12 (1981), 377–92

— 'Aspects of Modelling and Simulating Tumour Growth and Treatment', *Journal of Cancer Research and Clinical Oncology*, 105 (1983), 1–12

Ede, D.A. and Law, J.T. 'Computer Simulation of Vertebrate Limb Morphogenesis', *Nature*, 221 (1969), 244–8

Goodwin, B.C. and Briere, C. (1985) 'Morphogenesis in *Acetabularia*'. In preparation

Maruyama, M. 'The Second Cybernetics: Deviation-amplifying Mutual Causal Processes', *American Scientist*, 51 (1963), 164–79

Matela, R.J. and Fletterick, R.J. 'A Topological Model of Cellular Self-sorting', *Journal of Theoretical Biology* 76 (1979), 403

— 'Computer Simulation of Cellular Self-sorting: A Topological Exchange Model', *Journal of Theoretical Biology*, 84 (1980), 673–90

Matela, R.J. and Ransom, R. 'A Topological Model of Cell Division: Structure of the Computer Program', BioSystems, 18(1) (1985), 65–78

— and Bowles, M.A. 'Computer Simulation of Compartment Maintenance in the *Drosophila* Wing Imaginal Disk', *Journal of Theoretical Biology*, 103 (1983), 357–78

Meinhardt, H. 'Morphogenesis of Lines and Nets', *Differentiation* 6 (1976), 117–23

— 'Cell Determination Boundaries as Organizing Regions for Secondary Embryonic Fields', *Developmental Biology*, 96 (1983), 375–85

Ransom, R. (1975) 'Computer Analysis of Division Patterns in the *Drosophila* Head Disk. *Journal of Theoretical Biology*, 53 (1975), 445–62

— 'Computer Analysis of Cell Division in *Drosophila* Imaginal Disks: Model Revision and Extension to Simulate Leg Disk Growth', *Journal of Theoretical Biology*, 66 (1977), 361–77

— *Transcription and Translation: Programs for the Apple II and IBM PC* (Elsevier Biosoft, Cambridge, 1986)

— and Matela, R.J. 'Computer Modelling of Cell Division During Development: A Topological Approach. *Journal of Embryology and Experimental Morphology*, Suppl. 83 (1984) 233–59

Spain, J.D. *BASIC Microcomputer Models in Biology* (Addison-Wesley, Reading, 1982)

Ulam, S. 'On Some Mathematical Problems Connected with Patterns on Growth Figures', *Proceedings of the Symposium on Applied Mathematics*, 14 (1962), 215–24

Von Neumann, J. (edited by W. Burks), *Theories of Self-Replicating Automata* (University of Illinois Press, Chicago, 1953)

Walton, J.S. and Risen, W.M.Jr 'Computer Animation: On-line Dynamic Display in Real Time', *Journal of Chemical Education* 334 (1969)

Wein, M. and Burtnyk, N. 'Computer Animation', in J. Belzar, A.G. Holtzman and A. Kent (eds) *Encyclopedia of Computer Science and Technology* 5, 397 (Marcel Dekker, New York, 1976)

Appendix 1: Matrix Operations

1.1 BASIC DEFINITIONS

The following definitions are due to Campbell (1968). A formal treatment of matrix theory is given in Gantmacher (1960).

Definition 1

A matrix **A** is a rectangular array of elements and can be denoted by:

$$\mathbf{A} = \begin{bmatrix} a_{11} & a_{12} & \cdots & a_{1n} \\ a_{21} & a_{22} & \cdots & a_{2n} \\ \cdot & \cdot & \cdots & \cdot \\ \cdot & \cdot & \cdots & \cdot \\ \cdot & \cdot & \cdots & \cdot \\ a_{m1} & a_{m2} & \cdots & a_{mn} \end{bmatrix}$$

For our purposes, the elements of a matrix will be assumed to belong to the field of complex numbers, but in reality they could be any element of a specified algebraic system. We shall refer to such elements as scalars.

Example 1.

$$\begin{bmatrix} 1 & 3 \\ 0 & 1 \end{bmatrix} \qquad [3/4 \quad 1 \quad 7] \qquad \begin{bmatrix} 2 & 0 & i \\ 3 & 7/3 & i \end{bmatrix}$$

A matrix is called a zero or null matrix if all the elements are zero. This matrix is denoted by 0. The matrix **A**, as given in definition 1 consists of m rows and n columns. The subscript i for the element a_{ij} designates the row in which the element occurs, and the subscript j designates the column. The double subscript ij is termed the address of the element. The order of a matrix is given by the number of rows (always stated first) followed by the number of columns. For example, if $m = 3$ and $n = 2$, we would say this matrix has 'order 3 by 2'. If $m = n$, the matrix is square and order n. The main diagonal of a square matrix are those elements where $i = j$.

Example 2.

$$\mathbf{A} = \begin{bmatrix} 1 & 5 & 2 \\ 3/2 & 0 & 1 \\ 3 & 5/3 & 7 \end{bmatrix}$$

is a square, real matrix of order 3. The main diagonal elements are 1, 0, 7. The element a_{23} is 1 (second row, third column).

Definition 2

If some rows and columns (or both) of a matrix **A** are deleted, the remaining array of elements is called a submatrix of **A**. By convention, **A** is a submatrix of itself.

Example 3.

$$\mathbf{A} = \begin{bmatrix} 1 & 2 & 3 \\ 4 & 5 & 6 \end{bmatrix}$$

We can define the following submatrices:

$$\begin{bmatrix} 1 & 2 & 3 \\ 4 & 5 & 6 \end{bmatrix} \quad \begin{bmatrix} 1 & 2 \\ 4 & 5 \end{bmatrix} \quad \begin{bmatrix} 1 & 3 \\ 4 & 6 \end{bmatrix} \quad \begin{bmatrix} 2 & 3 \\ 5 & 6 \end{bmatrix}$$

$$\begin{bmatrix} 1 \\ 4 \end{bmatrix} \quad \begin{bmatrix} 2 \\ 5 \end{bmatrix} \quad \begin{bmatrix} 3 \\ 6 \end{bmatrix} \quad [1 \;\; 2 \;\; 3], \quad [4 \;\; 5 \;\; 6]$$

$$[1], [2], [3], [4], [5], [6], [1 \;\; 2], [1 \;\; 3], [4 \;\; 5], [4 \;\; 6]$$

$$[2 \;\; 3], [5 \;\; 6]$$

This set contains all the submatrices for the matrix **A**. An abbreviated notation for expressing a matrix is given by

$$\mathbf{A} = [a_{ij}] \; (m,n)$$

where i varies from 1 to m and j from 1 to n.

There are occasions when it is convenient to divide the matrix into submatrices by means of dashed lines. For example, if matrix **A** has order 3 × 3 [square matrix] then

$$\mathbf{A} = \begin{bmatrix} \mathbf{A}_{11} & \vdots & \mathbf{A}_{12} \\ \cdots & \vdots & \cdots \\ \mathbf{A}_{21} & \vdots & \mathbf{A}_{22} \end{bmatrix} = \begin{bmatrix} a_{11} & a_{12} & \vdots & a_{13} \\ a_{21} & a_{22} & \vdots & a_{23} \\ \cdots & \cdots & \cdots & \cdots \\ a_{31} & a_{32} & \vdots & a_{33} \end{bmatrix}$$

would be a partition of **A**.

1.2 VECTORS

Matrices of order 1 by n are termed row matrices, and matrices of order n by 1 are called column matrices.

Definition 3

A vector $\mathbf{v}$ of order n is an ordered set of n scalars $(a_1, a_2, \ldots, a_n)$.

Example 4. The coordinates of a point $(x \;\; y \;\; z)$ in three-dimensional space are an ordered set of scalars and can be considered as a vector. The set is ordered because, by convention, x precedes y which precedes z.

Definition 4

Two vectors of the same order may be added by adding the correspondent elements.

Let $\mathbf{A} = [a_1 \quad a_2 \quad a_3]$ and $\mathbf{B} = [b_1 \quad b_2 \quad b_3]$ then $\mathbf{A} + \mathbf{B} = [a_1 + b_2 \quad a_2 + b_2 \quad a_3 + b_3]$

Example 5.

$$\mathbf{A} = [1 \;\; 2 \;\; 3], \quad \mathbf{B} = [1 \;\; 6 \;\; 4]$$
$$\mathbf{A} + \mathbf{B} = [2 \;\; 8 \;\; 7]$$

Definition 5

A vector is multiplied by a scalar by multiplying each element of the vector by the given scalar. Let $\mathbf{A} = [a_1 \quad a_2 \quad a_3]$ and s be a scalar then $s\mathbf{A} = [sa_1 \quad sa_2 \quad sa_3]$

Example 6.

$$\mathbf{A} = [1 \quad 2 \quad 3], \; s = 3$$
$$s\mathbf{A} = [3 \quad 6 \quad 9]$$

Definition 6

Let $\mathbf{A} = [a_1 \; a_2 \; \ldots \; a_n]$ and $\mathbf{B} = [b_1 \; b_2 \; \ldots \; b_n]$ be two real vectors of the same order. The dot product (inner product or scalar product) of **A** and **B**, denoted by $\mathbf{A} \bullet \mathbf{B}$ is:

$$\mathbf{A} \bullet \mathbf{B} = a_1 b_1 + a_2 b_2 + \ldots + a_n b_n$$

The definitions given in (4) and (5) are used to define subtraction of the vector as

$$\mathbf{A} - \mathbf{B} = \mathbf{A} + (-1)\mathbf{B}$$

1.3 MATRIX ADDITION

Definition 7

Given two matrices $\mathbf{A} = [a_{ij}](m,n)$ and $\mathbf{B} = [b_{ij}](m,n)$ their sum is defined as $\mathbf{A} + \mathbf{B} = [a_{ij} + b_{ij}](m,n)$. Two matrices are said to be conformable for addition if they are of the same order.

Example 7.

$$\mathbf{A} = \begin{bmatrix} 1 & 2 \\ 3 & 4 \end{bmatrix} \quad \mathbf{B} = \begin{bmatrix} 1 & 0 \\ 2 & 3 \end{bmatrix}$$

$$\mathbf{A} + \mathbf{B} = \begin{bmatrix} (1+1) & (2+0) \\ (3+2) & (4+3) \end{bmatrix} = \begin{bmatrix} 2 & 2 \\ 5 & 7 \end{bmatrix}$$

The following theorems are stated without proof

Theorem 1: Commutative. If two matrices **A** and **B** with scalar entries are conformable then $\mathbf{A} + \mathbf{B} = \mathbf{B} + \mathbf{A}$.

Theorem 2: Associative Property. If three matrices **A**, **B**, **C** with scalar entries are of the same order then $\mathbf{A} + (\mathbf{B} + \mathbf{C}) = (\mathbf{A} + \mathbf{B}) + \mathbf{C}$.

Theorem 3: Cancellation Property. If three matrices **A**, **B**, **C** with scalar entries are of the same order then $\mathbf{A} + \mathbf{B} = \mathbf{A} + \mathbf{C} \Rightarrow \mathbf{B} = \mathbf{C}$ and $\mathbf{A} = \mathbf{B} \Rightarrow \mathbf{A} = \mathbf{C}$.

1.4 THE TRACE OF A MATRIX

A very useful unary operation which has some interesting properties is the *trace* of a matrix. We note that the result of this unary operation on a square matrix with scalar entries is itself a scalar.

Definition 8

The trace of a square matrix **A** is the sum of the elements on the main diagonal and is denoted as Tr **A**.

Example 8.

$$\text{Tr}\begin{bmatrix} a_{11} & a_{12} & a_{13} \\ a_{21} & a_{22} & a_{23} \\ a_{31} & a_{32} & a_{33} \end{bmatrix} = \begin{bmatrix} a_{11} + a_{22} + a_{33} \end{bmatrix}$$

$$\text{Tr}\begin{bmatrix} 1 & 2 & 3 \\ 0 & 5 & 3 \\ 4 & 7 & 8 \end{bmatrix} = \begin{bmatrix} 1+5+8 \end{bmatrix} = 14$$

The following theorems relating to the trace of a matrix are stated without proof.

Theorem 4. If $\mathbf{A}$ is a square matrix, then Tr $\mathbf{A}$ = Tr $\mathbf{A}^T$ where $\mathbf{A}^T$ is the transpose of matrix $\mathbf{A}$. That is, row i is replaced with the column i.

Theorem 5. If $\mathbf{A}$ and $\mathbf{B}$ are $n \times n$ matrices, then Tr $(\mathbf{A} + \mathbf{B}) =$ Tr $\mathbf{A}$ + Tr $\mathbf{B}$.

Theorem 6. If $\mathbf{A}$ is an $n \times n$ matrix and c is a scalar, then $\text{Tr}(c\mathbf{A}) = c\text{Tr}\mathbf{A}$.

Theorem 7. If $\mathbf{A}$ and $\mathbf{B}$ are $n \times n$ matrices and $\mathbf{AB} = \mathbf{BA}$ then $\text{Tr}(\mathbf{AB}) = \text{Tr}(\mathbf{BA})$ (*Note* $\mathbf{AB} \neq \mathbf{BA}$ usually, for example.

if $\mathbf{A} = \begin{pmatrix} a & b \\ c & d \end{pmatrix}$ and $\mathbf{B} = \begin{pmatrix} x & y \\ z & w \end{pmatrix}$

then $\mathbf{AB} = \mathbf{BA}$ if and only if bz = cy).

1.5 THE DETERMINANT OF A MATRIX

We now describe a most important unary operation on a square matrix, termed a determinant and denoted as det $\mathbf{A}$ or $\mathbf{A}$.

Definition 9

If the ith row and jth column of an $n \times n$ matrix $\mathbf{A}$ are deleted ($n \geqslant 2$), the determinant of the resulting submatrix is called the minor of a_{ij} and is denoted $\mathbf{M}ij$. The quantity $\mathbf{A}ij = (-1)\ \mathbf{M}ij$ is called the cofactor of a_{ij}.

Example 9. Consider the matrix $\mathbf{A}$ as defined by

$$\mathbf{A} = \begin{bmatrix} a_{11} & a_{12} & a_{13} \\ a_{21} & a_{22} & a_{23} \\ a_{31} & a_{32} & a_{33} \end{bmatrix}$$

The minor of a_{21} is given by

$$\mathbf{M}_{21} = \det \begin{bmatrix} a_{12} & a_{13} \\ a_{32} & a_{33} \end{bmatrix}$$

and the cofactor of a_{21} is

$$\begin{aligned} \mathbf{A}_{21} &= (-1)^{2+1}\,\mathbf{M}_{21} \\ &= (-1) \det \begin{bmatrix} a_{12} & a_{13} \\ a_{32} & a_{33} \end{bmatrix} \end{aligned}$$

Definition 10

If a matrix contains a single entry, then the determinant of the matrix is equal to that entry. If $\mathbf{A}$ is a square matrix of order n, ($n \geqslant 2$), where i and j are fixed integers such that $1 \leqslant i \leqslant n$ and $1 \leqslant j \leqslant n$, then its determinant is:

$$\begin{aligned} \det \mathbf{A} &= a_{i1}\,\mathbf{A}_{i1} + a_{i2}\,\mathbf{A}_{i2} + \ldots + a_{in}\,\mathbf{A}_{in} \\ &= a_{ij}\,\mathbf{A}_{ij} + a_{2j}\,\mathbf{A}_{2j} + \ldots\ a_{nj}\,\mathbf{A}_{nj}. \end{aligned}$$

The following theorem is stated without proof:

Theorem 8. The value of det $\mathbf{A}$ is the same no matter which row i or column j is chosen.

1.6 MULTIPLICATION BY A SCALAR

Definition 11

Given a matrix $\mathbf{A}\,[a_{ij}](m,n)$ and a scalar s, then $s\mathbf{A} = [sa_{ij}](m,n)$ and $\mathbf{A}s = [a_{ij}s](m,n)$. That is, a matrix is multiplied by a scalar by multiplying every element of the matrix by that scalar.

Example 10.

$$\text{Let } \mathbf{A} = \begin{bmatrix} 1 & 2 \\ 3 & 4 \end{bmatrix} \text{ and } s = 3$$

then

$$s\mathbf{A} = \mathbf{A}s = 3 \begin{bmatrix} 1 & 2 \\ 3 & 4 \end{bmatrix} = \begin{bmatrix} 3 & 6 \\ 9 & 12 \end{bmatrix}$$

1.7 MATRIX MULTIPLICATION

The final binary operation that we shall consider is matrix multiplication. Consider the product of a row vector of n coefficients and a column vector of n variables. The result of this multiplication is a 1×1 matrix whose single entry is a linear function of the n variables. For multiplication to be performed the matrices must be conformable. For example, say we wish to multiply a 'n by m' matrix by a 'p by q' matrix. These matrices must conform in that 'm must equal p', and the dimensions or order of the resulting matrix is given as 'n by q'.

Example 11.

(1) 2 by 3 and 2 by 3, result: non-conformable.
(2) 2 by 3 and 3 by 2, result: 2 by 2 matrix.

Definition 12

Let **A** be an n by m matrix and **B** be a p by q matrix. Remember that m must equal p for multiplication. The product **C** = **AB** is an n by q matrix, where each entry c_{ij} of **C** is obtained by multiplying corresponding entries of the ith row of **A** by those of the jth column of **B** and adding the results. If **A** and **B** are real matrices, then each c_{ij} entry is simply the dot product of the ith row of **A** and jth column of **B**. This is illustrated in the following diagram:

$$\text{If } \mathbf{A} = \begin{bmatrix} a_{11} & a_{12} & \cdots & a_{1m} \\ a_{21} & a_{22} & \cdots & a_{2m} \\ \cdot & \cdot & \cdots & \cdot \\ \cdot & \cdot & \cdots & \cdot \\ \cdot & \cdot & \cdots & \cdot \\ a_{n1} & a_{n2} & \cdots & a_{nm} \end{bmatrix}$$

$$\text{and } \mathbf{B} = \begin{bmatrix} b_{11} & b_{12} & \cdots & b_{1q} \\ b_{21} & b_{22} & \cdots & b_{2q} \\ \cdot & \cdot & \cdots & \cdot \\ \cdot & \cdot & \cdots & \cdot \\ \cdot & - & \cdots & \cdot \\ b_{m} & b_{m2} & \cdots & b_{mq} \end{bmatrix}$$

$$\text{then } \mathbf{C} = \mathbf{AB} = \begin{bmatrix} c_{11} & c_{12} & \cdots & c_{1q} \\ c_{21} & c_{22} & \cdots & c_{2q} \\ \cdot & \cdot & \cdots & \cdot \\ \cdot & \cdot & \cdots & \cdot \\ \cdot & \cdot & \cdots & \cdot \\ c_{n1} & c_{n2} & \cdots & c_{nq} \end{bmatrix}$$

where $c_1 = a_{11}b_{11} + a_{12}b_{21} + \ldots + a_{1m}m_{m1}$.

Example 12:

$$\text{Let } \mathbf{A} = \begin{bmatrix} 1 & 2 \\ 1 & 2 \end{bmatrix} \quad n=2, m=2$$

$$\mathbf{B} = \begin{bmatrix} 2 & 3 & 0 \\ 1 & 2 & 3 \end{bmatrix} \quad p=2, q=3$$

$$\text{then } \mathbf{AB} = \begin{bmatrix} (1 \times 2+1 \times 1) & (1 \times 3+1 \times 2) & (1 \times 0+1 \times 3) \\ (1 \times 2+2 \times 1) & (1 \times 3+2 \times 2) & (1 \times 0+2 \times 3) \end{bmatrix}$$

$$= \begin{bmatrix} 3 & 5 & 3 \\ 4 & 7 & 6 \end{bmatrix}$$

1.8 REFERENCES

Campbell, H.G. Matrices with Applications (Appleton-Century-Crofts, Education Division, Meredith Corporation, New York, 1968)

Gantmacher, F.R. *The Theory of Matrices*, vol 1 (Chelsea Publishing Company, New York, 1960)

Appendix 2: A Graphics Glossary

The computer graphics literature contains a large number of terms that are unfamiliar to the non-expert computer user. Some terms that are used outside of the graphics area also have very specific meanings in computer graphics. The following definitions are based on the fuller listing given in *Computer Graphics* 15 (2) (1981). Hardware definitions have been kept to a minimum as these are covered in detail in Chapter 2.

Absolute point An individually addressable position on the display surface marked by given *x* and *y* coordinates.

Addressable position The smallest discrete unit in the viewing area to which a CRT beam or plotter pen may be moved.

Aliasing A defect in the image caused by improper sampling on a raster display. Typical aliasing effects are 'staircase effect' lines, and Moire patterns.

Animation The process of presenting a series of images in rapid succession to create the illusion of movement.

Area filling The process of colouring or patterning pixels in a given area.

Aspect ratio The height-to-width ratio of a display view area.

Attribute A characteristic of a primitive, for example colour, intensity level or line type.

Beam current The rate of flow of electrons from an electron gun.

Bitmap An area of digital memory containing a description of the state of addressable pixels in a raster display.

Character generator Implementation of the dot or line patterns representing characters in ROM, with subsequent display to form a character on the screen.

Clipping A process to ensure that a displayed image is not drawn beyond predetermined boundaries.

Coherence The inherent property of objects in a raster display where neighbouring pixels (or identical pixels in adjacent frames) tend to possess similar attributes. It is used by scan line algorithms to increase efficiency.

Colour look-up table An array of colour values defined by their red, green and blue components, stored externally to the bitmap.

Coordinates An ordered set of data values, either absolute or relative, specifying a location in space.

Core graphics system A standard device independent subroutine package established by ACM SIGGRAPH for interactive display devices.

Cursor A special symbol used to highlight a particular position on the screen. The cursor normally indicates the position for display of the next alphanumeric character or for input of data.

Depth cueing A method of simulating depth in a three-dimensional image. Possible techniques are modulation of line intensity, perspective, and stereoscopic effects.

Device driver The software for conversion of device-independent graphics commands into device-specific display commands.

Device space The area defined by the coordinate system of the display surface.

Digitiser A device for coding images or shapes into numeric data.

Display address space The area defined by the set of display coordinates.

Display buffer A storage device or memory area holding all display orders and coordinate data required to generate an image.

DMA Direct memory access. A method of transferring display data to and from host computer memory without processing by the host CPU.

Erase The process of removal of display items from the display surface.

Fill See: Area filling.

Flicker An undesired blinking or pulsation of a CRT when the refresh rate is too low.

Function generator A hardware unit used to control CRT beam movements or intensity and to generate required functional information on the screen from a specific input definition.

Graphic language A language used to program a display device.

Grid Two mutually orthogonal sets of parallel lines.

Hardcopy A permanent copy of the information on a display that can be separated from the display.

Hidden lines Line segments that should be obscured from view in a projected image of a three-dimensional object. These lines may be eliminated or represented as lines with different textures or colour intensities.

Highlight A display characteristic intended to draw attention to a display item.

Icon A symbol in a menu used to represent a command.

Image A collection of displayed items.

Image space The area defined by the coordinate system of the view plane.

Interactive mode A method of operation that allows online machine–man communciation.

Interlace scan A raster display technique to reduce flickering. The beam scans the even lines first and then scans the odd lines.

Line style One of the attributes of a visible line: for example, solid, dashed, dotted.

Mapping function A transformation which converts the elements of one representational system into another: for example, model space to image space, or window to viewport.

Mesh Two sets of arbitrary parallel curves.

Metafile A collection of commands for an output device used in some implementations of the Core System. A device-independent intermediate display file used to communicate pictures between users and computers.

Model space See: World coordinates.

Origin A reference point with zero coordinates.

Output primitive A type of picture element: for example line, text, polyline or marker.

Parallax Displacement of an object seen from two different points, for instance a point as seen by the light pen or the eye.

Parallel projection Projection of a point in three-dimensional space onto the view plane by a perpendicular line.

Persistence The decay period of an image on the phosphor of a CRT.

Perspective projection Projection of a point in three-dimensional space onto the view plane with depth information — distant objects appear smaller than closer ones.

Pixel A single picture element; the smallest displayable area on the raster display surface.

Plotter step size The horizontal or vertical distance between two adjacent addressable points on a plotter (compare: Pixel).

Raster scan Image generation using an intensity controlled line-by-line sweep of the display surface.

Raster unit The vertical or horizontal distance between two adjacent addressable points on a CRT.

Reflectance A surface property of a three-dimensional object determining the amount of incident light reflected.

Refresh rate The rate at which a display is regenerated.

Resolution The smallest distance between two display elements observable by a viewer.

Rotate To move a display item about an axis in the image space.

Scaling A transformation function altering a display item by multiplying its coordinates by constant values.

Scan conversion The definition of an image as a series of points on a group of raster lines.

Shading Computation of the colours and intensities of the surface of a three-dimensional object.

Text string A string of consecutive text characters for display purposes.

Translation The movement of a display item from one point to another in the image space without rotation.

Vector A graphic element posessing magnitude and direction.

Viewing transformation Mapping of positions in world coordinates to positions in device coordinates.
View plane Representation of a surface in two dimensions.
Viewport A bounded area within device space on which a window is displayed.
Window A bounded area within image space.
World coordinates The model coordinate system.
Zoom The process of scaling all elements of a viewport to give the appearance of movement away from or towards the observer.

Index

COMPUTER GRAPHICS IN BIOLOGY

ADVANCES IN PLANT SCIENCES – VOL. I
General Editor: T.R. Dudley, Ph.d.

CHICAGO PUBLIC LIBRARY
R00660 31857